Allan Greenwell, W. T. Curry

Rural Water Supply

A Practical Handbook on the Supply of Water & Construction of Waterworks for

Small Country Districts

Allan Greenwell, W. T. Curry

Rural Water Supply
A Practical Handbook on the Supply of Water & Construction of Waterworks for Small Country Districts

ISBN/EAN: 9783337139179

Printed in Europe, USA, Canada, Australia, Japan

Cover: Foto ©Andreas Hilbeck / pixelio.de

More available books at **www.hansebooks.com**

Rural Water Supply

A PRACTICAL HANDBOOK

ON THE

SUPPLY OF WATER AND CONSTRUCTION OF WATERWORKS FOR SMALL COUNTRY DISTRICTS

BY

ALLAN GREENWELL, A.M.I.C.E.

AND

W. T. CURRY, A.M.I.C.E., F.G.S.

LONDON
CROSBY LOCKWOOD AND SON
7 STATIONERS' HALL COURT, LUDGATE HILL
1896

PREFACE BY G. C. GREENWELL,

M.I.C.E., F.G.S.

———◆◇◆———

ONE of the greatest necessities of human life is an ample supply of wholesome water; and few objects are occasionally more difficult of attainment, not so much on account of any approach to impracticability as of certain hindrances to the mode of setting about it.

The situation of individuals with respect to each other renders that combination of the constituent parts of society, which is necessary in order to bring about a general co-operation, very difficult to adjust.

The supply of water enjoyed by one portion of the community may be sufficient both in quantity and in quality for all of its requirements; but the less happily circumstanced remainder may be dangerously near want. On the one hand, water is demanded, even though at the *cost* of the whole community; while on the other, such compulsion appears to be extremely unjust.

We are thus, at the very outset of the question of coercive water supply, met by antagonisms, which most certainly have not as yet been reconciled, with the result that many desirable schemes have either been quashed in their incipience, or, if completed under pressure, have given rise to heartburnings which have been productive of anything but peace and goodwill.

This early stage of the proceedings is, however, in the

present work supposed to have been passed over: the "tempest in the teapot" has exhausted itself, the strong will has prevailed, and the weak has gone to the wall as usual; the powers necessary to establish the waterworks have been obtained, and to direct in plain language how the precious element is to be brought from its hidden spring or other sources for the good of the population of the rural districts is the object of this little work.

There are various treatises, such as those by Messrs. Humber, Burton, Turner and Brightmore, and Fanning, as well as innumerable papers, which admirably deal with the questions of Water Engineering; but most, if not all of these, are of a more or less elaborate character; and it appeared to the writers that an elementary work on the subject was much to be desired—such as should enable the student to acquire a knowledge of the principles and construction of waterworks, simple in detail and efficient in application; which would qualify him to arrange and complete a system of waterworks on a moderate and effective scale—due qualification in which respect would lay the foundation for the future mastery of more important schemes.

The Authors have endeavoured constantly to keep in view, so far as lay in their power, simplicity of expression and of formulæ, in the hope of producing such a book as should be readable and intelligible by every one who earnestly desires to step on the lower rung of the ladder with the determination to climb gradually to the top.

G. C. GREENWELL.

Duffield, near Derby,
May, 1895.

AUTHORS' PREFACE.

THIS little contribution to the literature of Waterworks Engineering is based upon a series of articles which appeared in the "Student's Column" of the *Builder*, from July to December, 1894, the articles having since been revised and brought up to date.

The Authors desire to express their grateful acknowledgment for the valuable information which they have obtained from the many standard works referred to in the text.

A. G.

W. T. C.

LONDON,
    September, 1895.

CONTENTS.

———⋄———

RURAL WATER SUPPLY.

CHAPTER I.

PRELIMINARY.

THE benefit of a plentiful supply of wholesome water in the country is hardly to be overestimated. Until recent years general systems of supply were almost entirely confined to towns, and, except in a few cases, where liberal-minded landowners carried out gravitation, or even small pumping schemes for the supply of water to their estates, rural populations were largely dependent upon open streams and shallow wells, which are always liable to pollution. The necessity for a proper supply of water to towns and congested populations generally is evident to all; but the idea that there should be any difficulty in procuring potable water in the country would seem incredible to most people who had not made rural life more or less of a study. The matter, however, rises to great importance when we remember that country areas are the source of supply of most of the food which is consumed in towns; and that when, for instance, a farm supplies milk for sale, any evil conditions under which that milk is produced follow it; and an impure water supply at a farm is frequently the cause of an outbreak of typhoid fever amongst the consumers of the milk at a distance.

A great change has, however, recently taken place, and

B

schemes of rural water supply, especially in the Midlands, have become as plentiful as blackberries. This is partly attributable to the advance in sanitary knowledge, and to the necessities of rural populations becoming daily more pressing.

Sources liable to pollution tend to become more polluted, and pollution, generally speaking, is cumulative. The recent exceptionally dry seasons have brought to light distressing cases of privation, and one exposure has led to others.

The Local Government Act, 1888, has enabled County Councils to put considerable pressure upon local authorities, when they are of opinon, on the report of their Medical Officer of Health, that such authorities are not properly performing their duties under the Public Health Acts, one of which is to provide efficient water supplies for their districts.

The Local Government Act, 1894 (commonly known as the Parish Councils Act), includes two sections, which will probably facilitate the provision of many rural water supplies, which have hitherto appeared impracticable.

Sec. 8 gives power to a parish council "to utilize any well, spring, or stream within their parish, and providing facilities for obtaining water therefrom, but so as not to interfere with the rights of any corporation or person."

Sec. 16 provides that where a parish council resolve that a rural district council ought to have provided the parish with a supply of water " in cases where danger arises to the health of the inhabitants from the insufficiency or unwhole-someness of the existing supply of water " (Public Health Act, 1875, sec. 299), complaint may be made to the County Council, which may transfer those duties to itself, or may appoint a person to perform the duty.

There are very important differences between urban and rural water supply, which often make the difficulties to be contended with much greater in the latter than in the former. Except where the scheme is carried out by private indivi-duals for the benefit of their estates, the duty of providing an efficient water supply usually devolves upon the local authority.

By the Public Health Act, 1875, sec. 299, it becomes the duty of a Local Authority to provide their district with a supply of water, "in cases where danger arises to the health of the inhabitants from the insufficiency or unwholesomeness of the existing supply of water, and a proper supply can be got at a reasonable cost." The money for carrying out such a supply is usually obtained on loan (after a Local Government Board inquiry), repayable by yearly or half-yearly instalments of principal and interest in thirty years. The present rate of interest is usually three and a half per cent. per annum, which makes the annual instalment £5 8s. 9d. per cent.

This annual charge, together with the working expenses, has to be met either by a water-rate over the whole area, or by charges made upon the consumers, or by both. In an urban district, where nearly the whole population benefits by the supply, a water-rate is not necessarily a hardship; but in rural districts, where the area is determined by parish boundaries, and only a small portion of the population benefits by the supply, the case is different. As a consequence of the shuffling which frequently becomes possible through the intricacies of the law upon the subject, the few are sometimes benefited at the expense of the many.

A small village had a fairly good water supply, but it included a patch of elevated ground excellently suited for building sites, but where water was conspicuous by its absence. Largely through the instrumentality of a local architect, himself a leading member of the local sanitary authority, this patch of ground was covered with houses of the villa class, in spite of section 6 of the Public Health (Water) Amendment Act, 1878, which makes it unlawful in any rural district for the owner of any dwelling-house which may be erected or rebuilt after the passing of this Act, to occupy or permit the same to be occupied without first obtaining the certificate of the sanitary authority, "that there is provided, within a reasonable distance of the house, such an available supply of wholesome water as may appear to such authority to be sufficient for the consumption and use for domestic purposes of the inmates of the house."

The houses being built and occupied, it became necessary to provide them with water; so a scheme was approved and carried out by the Local Sanitary Authority. To meet the expenditure attendant upon the first cost of the scheme, for obvious reasons, a water-rate was made over the whole parish, instead of a charge being made on the consumers only. One of the inhabitants of the village, who resided upon his own property, had, previous to the above events, expended over a hundred pounds in order to efficiently supply his house with water from an excellent, never-failing well upon his premises. Notwithstanding that this supply had been pronounced of exceptional purity by competent analysts, this unfortunate owner was compelled to contribute some £5 a year for a commodity which he in no way required. It is needless to observe that this individual was not the only sufferer.

In certain cases the Local Government Board will consent to a special district being constituted, excluding as far as possible such areas as will not receive or do not require a supply. The Local Government Board, however, are rarely in favour of this step for purposes of water supply alone. The only alternative, therefore, is to make a charge upon the consumers sufficient to cover the periodical instalments of principal and interest, as well as the working expenses of the scheme.

The maximum charge which may lawfully be made (except under special circumstances) where a house is without a proper supply of water, and where a supply is enforced by a local authority, is 2*d.* a week per house, or 8*s.* 8*d.* a year. Where, however, the supply is given by agreement, the authority may make such reasonable charge as they think fit. Where the rate is levied upon the consumers only, it must be so adjusted as not to produce a profit which would benefit the ratepayers at large at the expense of the consumers.

A usual charge is 2*d.* per week for houses with a rateable value under ten pounds per annum, and 5 per cent. per annum when the rateable value exceeds that amount. This scale averages 5 per cent. per annum throughout.

If the annual instalment of principal and interest, together with the working expenses, does not exceed 5 per cent. on the rateable value of the property supplied, the scheme can be made self-supporting, as the remainder of the parish or parishes, receiving no benefit from the supply, need not be made to contribute. If, however, the expenditure exceeds the receipts, the balance must be met by a special water-rate levied over the whole area, irrespective of benefit.

As the rateable value of rural is considerably less than that of urban districts, area for area, and as the length of pipe necessary for the supply of a given number of houses is many times greater, the first cost of a water supply scheme for a rural area must of necessity be made relatively small and the working expenses reduced to a minimum. To secure this end gravitation schemes are usually the only means which can be entertained, as the 5 per cent. above-mentioned rarely allows a sufficient margin for the working expenses of a pumping establishment. Occasionally a self-acting pumping system becomes possible where a fall of water can be utilized to work a water-wheel, turbine, or hydraulic ram.

In the following pages, it is proposed to deal with such schemes as are usually feasible in rural districts, taking the above remarks into consideration. The various systems for affording such supplies will be described in detail, the principles explained, the machinery and materials carefully described, and the necessary information supplied both for preparing and carrying out the schemes. Plans, sections, specifications, and estimates of cost will be given, and, where possible, detailed prices of materials and workmanship will be indicated.

CHAPTER II

VARIOUS METHODS OF SUPPLY. ADVANTAGES AND DISADVANTAGES OF EACH.

THE selection of a source from which to obtain a water supply for a rural district, is dependent on a variety of considerations, among which are the following :—

1. Purity of the supply.
2. Volume and permanency.
3. Elevation with regard to the district to be supplied.
4. Distance from the district to be supplied.
5. Nature of intervening ground.
6. Purchase of water rights and easements.

In the "Suggestions as to Water Supply," etc., issued by the Local Government Board, the various sources from which water is usually obtained for purposes of domestic supply are arranged as follows :—

From mountain ranges, which act as condensers.

From rivers and streams.

From natural springs.

From wells artificially formed.

From impounding reservoirs.

From a combination of two or more of these sources.

Impurities likely to be met with in a source of water supply are of two kinds—

1. Those that can be removed by inexpensive means —*e.g.* mechanical filtration.

2. Those which cannot be thus removed, or in the removal of which heavy expense would be incurred.

In the former class are included organic matter of vegetable origin in suspension—*e.g.* peat, also non-poisonous mineral substances, such as carbonate of lime.

In the latter class are included the products of decaying organic matter of animal origin, as well as actual organic life, or what is generally known as sewage contamination; also poisonous substances, such as lead, and other substances, such as common salt (NaCl) which becomes injurious when present in large quantities.

Undoubtedly the first necessity of a water supply for domestic purposes is purity, and this must be assured at the outset.

In taking samples of water for purposes of analysis, a perfectly clean stoppered Winchester quart bottle (holding about half a gallon) should be used. The bottle should have been previously washed out with a little strong sulphuric (H_2SO_4) or hydrochloric (HCl) acid, and then rinsed with frequent changes of pure water until the rinsings do not redden a piece of blue litmus paper. Before taking the sample, the bottle and the stopper should be thoroughly rinsed with the water to be analyzed, and should then be filled to the neck with the water, stoppered, sealed, and labelled on the spot, and, if possible, analyzed within forty-eight hours.

In submitting a sample of water for analysis, as much information as possible should be given as to the situation of the source from which it has been taken, both geologically and with regard to any possible causes of pollutions in the vicinity. It is only by reading the analysis of a sample of water in close conjunction with the most careful observation of the surroundings and conditions of the source from which it has been taken, that any reliable opinion can be formed as to the suitability or otherwise of the supply for domestic purposes.

The River Pollution Commissioners, in their sixth report, classified the various sources with regard to potability as follows:

Wholesome	1. Spring water.	Very palatable.
	2. Deep well water.	
	3. Upland surface water.	Moderately palatable.
Suspicious	4. Stored rainwater.	
	5. Surface water from cultivated land.	
Dangerous	6. River water to which sewage gains access.	Palatable.
	7. Shallow well water.	

Rainfall is, practically speaking, the ultimate source of all water supply, and the nearer the source the purer, though not necessarily the more palatable, the water. Rainfall is disposed of in three ways:—

1. A portion is again evaporated.

2. Another portion flows over the surface of the ground to form streams and rivers.

3. The remainder sinks into the ground, and forms the underground reservoirs in which wells are sunk, issuing again at the lowest lip as springs.

1. *Spring water.*

The water from deep-seated springs is usually organically pure, though frequently highly charged with mineral substances. Where, however, the outcrop of the water-bearing stratum, at the point where it yields the spring, is of large area, and upon it houses, farmyards, and other possible sources of pollution are in existence, great care should be taken that the spring is not thereby affected.

A spring rises on the side of a hill at the junction of the upper green sand with the Oxford clay. A farmhouse is situated about 150 yards distant from the spring, and 100 feet above it, on the outcrop of the upper green sand. A sample of water from the spring was submitted to the county analyst for examination, and the following is an extract from his report thereon:—

" This water is plainly contaminated with the products of decomposition of animal matter, and is liable, as has been the case in several recorded instances, to carry the infective matter of specific disease. A consideration of the facts of the case confirms me in this opinion, and the amount of pollution is greater than at first sight appears, because the green sand furnishes a water of more than average purity: probably about one-half the solid matter of this water is directly derived from the farm sewage."

The intervening land between the farmyard and the spring is grass pasture, and there is no other discoverable source of pollution.

2. *Deep well water.*

Deep wells, especially those sunk through an impervious bed of considerable area, afford supplies of excellent quality. Care must be taken that the portion of the well above the impervious bed is so constructed as to prevent percolation from the surface or from the upper strata.

3. *Upland surface water.*

Water from this source is usually satisfactory, but is frequently discoloured with peat, even to such an extent as to render it unfit for domestic purposes.

4. *Stored rain water.*

Where there is no other available source, and there is freedom from smoke, etc., rain water may be used for domestic purposes, but it is unpalatable on account of the absence of aëration. It should be filtered before storage, and the tank should be well ventilated.

The three remaining sources are not fitted for domestic purposes. River water is, however, frequently used, but it requires efficient filtration, which renders it too expensive for use on a small scale.

The next points for consideration are the volume and permanency of the source. To obtain this information frequent gaugings, taken over a considerable period, are necessary. The area of the watershed relied upon, the greatest and least rainfall, percolation, and evaporation are all important factors of the result. These points will be considered in a future chapter. Information from the oldest inhabitants of the locality must not be allowed too much weight, or serious consequences may result.

In a recent survey for a village water supply, three

apparently deep-seated springs of most excellent quality were found in close proximity to each other. These springs were gauged on the 21st of March, 1893, and yielded a total volume of 820 gallons per minute. On April 8th they had fallen to 436 gallons per minute, and on May 6th to 207 gallons per minute. In June they were dry.

In the course of the same survey a spring was brought under notice, as to the permanency of which opinions were somewhat conflicting. Reference was made to the "oldest inhabitants," three nonogenarian labourers, who stated that they had known the spring all their lives, and that it had never yielded less than a certain flow. This was in February, 1893, and shortly after the spring failed.

The elevation of the source with regard to the district to be supplied is of great importance, for upon this will depend whether the natural flow of water by gravitation can be utilized, or whether pumping must be brought into requisition. Especially for a rural supply, gravitation should be secured if possible; for although the outlay is generally much greater, a heavy annual cost is avoided. Occasionally, where a sufficient quantity and fall of water are available, water-wheels, turbines, and hydraulic rams, which require no fuel and little attention, may be used; and in exposed situations, windmills are sometimes employed for pumping water on a small scale.

In deciding upon the relative merits of a gravitation and a pumping scheme, the distance of the source from the district to be supplied forms an important element; for if the distance over which the water has to be carried, in the former case, is great and the fall slight, pipes of large diameter may be required, entailing considerable expense, while in the latter case the length of pipe may be inconsiderable.

Especially in a gravitation scheme, on account of the relatively great distance over which the supply has usually to be carried, the question of easements is of great importance; whilst the question of water-rights affects both gravitation and pumping schemes to much the same extent.

These claims frequently lead to so great an expense as to necessitate the total abandonment of a water-supply scheme.

Every detail affecting those points should be carefully ascertained before any scheme is fully considered, and agreements *in writing* should be entered into with all parties in any way interested before works are commenced.

In a recent water-supply scheme, a claim was overlooked which might have been easily settled for £5. In consequence of this omission a trial took place four years later, costing the authority who carried out the scheme over £600.

CHAPTER III.

LAND VALUATION, RIPARIAN RIGHTS, EASEMENTS, AND COMPENSATION.

THE persons carrying out a water scheme are technically known as the "undertakers," and may be classified as follows :—

1. Private individuals or companies not possessing statutory or parliamentary powers.

2. Companies who have obtained a Provisional Order under the Gas and Water Facilities Acts, or the fuller powers of a private Act of Parliament.

3. Local Authorities, urban or rural, acting under

(a) The Public Health Acts;

(b) A Provisional Order;

(c) A Private Act of Parliament.

Private individuals or companies not possessing statutory or parliamentary powers are placed at a great disadvantage. They are liable to indictment or injunction for breaking up or obstructing the highways; they cannot acquire water or land, except by agreement; they cannot levy rates or make charges, except by agreement; and the only advantage which they can claim is that they "cannot be compelled to furnish a supply of water to any one on any terms."

It is obvious that, except where a landowner carries out a system of water supply for the benefit of his property, and where no difficulties arise as regards water-rights, easements, etc., or obstructive highway authorities, further powers are generally necessary.

The simplest and most economical way in which to obtain

such powers is, to obtain a Provisional Order under the Gas and Water Facilities Act, 1870, which must afterwards be confirmed by Parliament. A Provisional Order can only be obtained with the consent of the Local Sanitary Authority, and provided that there is not already in existence any legally empowered water company, able and willing to afford the necessary supply.

A Provisional Order (except in the case of a Local Sanitary Authority) does not confer compulsory powers as to purchase of land or water, or as to entry upon premises. To obtain these powers a special Act of Parliament is necessary.

The powers conferred by a special Act are very comprehensive, and are plainly set forth in the introduction to "The Law Relating to Gas and Water," by Messrs. Michael and Will, as follows :—

" Thus authorized by a special Act, a company may take compulsorily lands and streams, subject to the provisions and restrictions of the Lands Clauses Acts in exercising such powers. The undertakers must make to the owners and occupiers of, and all other parties interested in, any lands or streams taken or used for the purposes of the special Act, or injuriously affected by the construction or maintenance of the works thereby authorized, or otherwise by the execution of the powers thereby conferred, 'full compensation for the value of the lands and streams so taken or used, and for all damage sustained by such owners, occupiers, and other persons by reason of the exercise, as to such lands and streams, of the powers vested in the undertakers.' For the purpose of constructing waterworks, the undertakers may enter upon the lands and places described on the plans and in the books of reference, and may take the levels and set out the parts thereof, and dig and break up the soil, and trench and sough the same, and remove and use earth, stone, mines, minerals, trees, or other things. They may sink wells, make, maintain, alter, or discontinue reservoirs, waterworks, cisterns, tanks, aqueducts, drains, cuts, sluices, pipes, culverts, engines, and other works, and erect buildings ; they may also divert and impound water from the streams mentioned for that purpose in the special Act or the said plans or books of reference, and alter the course of such streams, not being navigable, and take such waters as may be found in and under, or on the lands to be taken for constructing the works. In the exercise of these powers, the undertakers are to do 'as little damage as can be.' With respect to the breaking up of streets for the purpose of laying pipes, the undertakers are empowered to open and break up the soil and pavement of the several streets and bridges within the limits of the special Act, and to open and break up sewers, drains, or tunnels, within or under the same, and to lay down pipes, conduits, service-pipes and other works and engines, and from time to time to repair, alter, or remove the same."

A Local Sanitary Authority (urban or rural) is placed in a somewhat different position to a company. Acting under the Public Health Act, 1875, a Local Sanitary Authority is invested with all the necessary powers for carrying out and afterwards maintaining a water-supply scheme, and in a pecuniary sense, is much better situated than a company, insomuch as it can compel the whole district to wholly or partially contribute the funds to meet the necessary expenditure, and is not restricted to the actual consumers of the water. Lands or easements necessary for any scheme of water supply must, however, be obtained by agreement; compulsory powers in this respect can only be secured by a Provisional Order or a special Act. A Local Authority having obtained a Provisional Order is, however, powerless, except by agreement, to purchase water-rights, *i.e.* the right to abstract water from streams, etc., and this difficulty can only be overcome by a special Act of Parliament.

A further difference existing between the powers of a local sanitary authority and those of a company, consists in the privilege which the former possesses of purchasing the easement only of laying pipes through land, whilst a landowner can compel a company to purchase the freehold of the land, except special provision has been made in the private Act.

As the supply of water to communities is in the majority of cases undertaken by the Local Sanitary Authority, the following pages will, unless otherwise stated, be specially adapted to assist in the preparation and execution of such schemes.

Sec. 308 of the Public Health Act, 1875, enacts as follows: "Where any person sustains any damage by reason of the exercise of any of the powers of this Act, in relation to any matter as to which he is not himself in default, full compensation shall be made to such person by the Local Authority exercising such powers; and any dispute as to the fact of damage or amount of compensation shall be settled by arbitration in manner provided by this Act, or if compensation claimed does not exceed the sum of twenty pounds, the same may, at the option of either party, be ascertained by, and recovered before, a court of summary jurisdiction."

Sec. 332 of the same Act further provides that: "Nothing in this Act shall be construed to authorize any Local Authority to injuriously affect any reservoir, canal, river, or stream, or the feeders thereof, or the supply, quality, or fall of water, contained in any reservoir, canal, river, stream, or in the feeders thereof, in cases where any body of persons or person would, if this Act had not passed, have been entitled by law to prevent or be relieved against the injuriously affecting such reservoir, canal, river, stream, feeders, or such supply, quality, or fall of water, unless the Local Authority first obtains the consent in writing of the body of persons or person so entitled as aforesaid."

These sections, taken together, indicate the importance attached to the purchase of and compensation for land and water rights, in connection with nearly every water-supply scheme. Unfortunately these matters are not regulated by recognized principles; but the following details, taken from actual experience, will act as a guide in most cases likely to occur in general practice.

In valuing property required for the construction of waterworks, the elements of value and interest to be

purchased are numerous, and vary according to the special circumstances of each case. The main points to be considered are: The value of the special adaptability of the site, of agricultural land, garden land, woods and plantations, houses and outbuildings, minerals, severance and injury, removal, trade and other fixtures, loss of trade profits, loss on forced sale.

Special Adaptability of the Site.

The " special adaptability " value has been brought into considerable prominence in waterworks cases during the last few years. This element of value was first urged in the case of Sir Walter Riddell and the Newcastle and Gateshead Water Company, and was afterwards an important feature in the case of the Countess Ossalinsky and the Corporation of Manchester, in reference to the Thirlmere scheme. These cases have been relied on as precedents in all the recent arbitrations under the Lands Clauses Act for waterworks purposes.

The valuation is made on the special adaptability of the land for reservoir or other purposes arising from the special and physical conditions of the property. Such value must not be based upon the value to the purchaser, but upon the value to the owner or seller.

In taking into consideration the special adaptability element, it is either taken as the whole value less the agricultural value, or in addition to the latter. In most cases, however, the agricultural value is excluded, and the property solely valued on its adaptability, to which must be added severance and other claims, if any. The value in such cases varies from £3 to £10 per acre, and at from twenty to twenty-eight years' purchase.

In the case of Riddell *v.* Newcastle and Gateshead Water Company, which went to the Court of Appeal, Lord Bramwell said as follows: "Special value in special circumstances should be adopted if you are dealing with reservoir sites, just as though you were dealing with building sites;

if you are wanting to buy land which is suitable for building purposes, you must pay building price for it."

Again, in the case of Ossalinsky *v.* Corporation of Manchester, the Court held, "if apart from the particular purchaser and the particular Act, land has enhanced value from any special circumstances, the owner is entitled to it." This case did not go to the Court of Appeal, the judgment being adopted by both parties.

Agricultural and Garden Land.

Freehold land is usually estimated on the nett annual rental, at from twenty-eight to thirty-five years' purchase.

Leasehold property is valued for the landlord on the rent and reversionary value, the tenants claiming the value of the unexpired term, with an allowance for forced sale and removal in proportion to the length of term.

The owner of a *reversion* is entitled to such a sum as would, if accumulated at interest until the date when the property will fall in, amount to the fee-simple value. The valuation is made on a basis of 3 per cent.

Woods and Plantations.

These are accurately measured for the acreage, and then arranged under the following heads:—

Full-grown timber.
Half-grown timber.
Young plantations.
Underwood.

Full-grown timber is in most cases measured, owing to the uncertainty of judging their value by the eye, allowance being made for bark and loss in cutting.

Half-grown timber is valued at an average price per acre, based upon the present age of the timber, and the time which will elapse before it will arrive at maturity. This period varies according to the situation, climate, and the nature of the trees.

C

Young plantations are valued on the outlay in producing them—viz. the plants, planting, fencing, draining, and other matters; and the rate of interest depending on the appearance and quality of the trees.

Underwoods are valued in a similar manner to half-grown timber.

The land occupied by the woods and plantations is estimated at the fee-simple value.

Houses and Outbuildings.

The value of these is estimated on the nett annual rental, less repairs, at twenty years' purchase, and an additional 10 per cent. for forced sale. Due regard must be paid, in making the valuation, to any prospective increase in the value of the property.

Minerals.

Minerals existing under lands to be valued, can only be properly estimated when such have been proved either on the land to be valued or on some adjoining property. The valuation is based upon the annual rental at from ten to sixteen years' purchase.

Severance and Injury.

When a portion only of a property is required by the undertakers, the remaining portion is frequently diminished in value, or even rendered practically useless, for sheep or stock farming or other purposes, owing to the want of shelter, water, accessibility, etc. These disadvantages must be estimated, and the owner and occupier compensated accordingly.

Removal.

This being in most cases a small matter, only a nominal compensation, if any, is allowed.

TRADE FIXTURES.

These are estimated according to the value of the machinery and other works, the horse-power (whether by steam, water, or horse-labour), the capability of the works for production or the ordinary work of the farm, and the state of repair.

LOSS OF TRADE PROFITS.

These are estimated according to the extent and nature of the business, whether the loss will be partial or entail the closing of an established business, and as to whether the occupier is owner, lessee, or tenant.

The value varies from one to six years' purchase of the profits, according to the circumstances.

LOSS FROM FORCED SALE.

This is, in some cases, considerable, but 10 per cent. is generally added as compensation for this loss.

CHAPTER IV.

LAND VALUATION, RIPARIAN RIGHTS, EASEMENTS, AND COMPENSATION—*continued*.

RIPARIAN RIGHTS AND COMPENSATION.

THE proprietor of any land adjoining or abutting on a stream has certain privileges or rights known as "riparian" (Lat. "ripa," a "river bank").

The law relating to these rights has been, from time to time, set forth in many well-known cases, to some of which further reference will be made.

In accordance with English law, the property in water flowing in a river or stream in its natural course belongs to no one, but the use of it to every one having a right of access to it.

In Miner *v.* Gilmour, Lord Kingsdown observed as follows: "By the general law applicable to running streams, every riparian proprietor has a right to what may be called the ordinary use of the water flowing past his land—for instance, the reasonable use of the water for his domestic purposes, and for his cattle, and this without regard to the effect which such use may have in case of deficiency upon proprietors lower down the stream. But, further, he has a right to the use of it for any purpose, or what may be termed the extraordinary use of it, provided that he does not thereby interfere with the rights of other proprietors, either above or below him. Subject to this condition, he may dam up for the purpose of a mill, or divert the water for the purpose of irrigation ; but he has no right to interrupt the regular flow of the stream, if he thereby interferes with the lawful use of the water by other proprietors, and inflicts upon them a sensible injury."

In the case of Chasemore *v.* Richards, it was held that "The right to the enjoyment of a natural stream of water on the surface belongs *naturali jure* to the proprietor of the adjoining land as a natural incidence to the right to the soil itself. He has the right to have it come to him in its natural state, in flow, quantity, and quality, and to go from him without obstruction, upon the same principle that he is entitled to the support of his neighbour's soil for his own in its natural state. And such a right depends in no way upon prescription, or the presumed grant of his neighbour, nor from the presumed acquiescence of the proprietors above and below."

Again, in Mason *v.* Hill, "A riparian proprietor can have no larger right than he has by nature against those above and below him. Hence the right to have a stream to flow in its natural state without diminution or alteration is an incident to the property in the land through which it passes; but flowing water is *publici juris*, not in the sense that it is a *bonum vacans*, to which the first occupant may acquire an exclusive right, but that it is public and common in this sense only, that all may reasonably use it who have a right of access to it; that none can have any property in the water itself, except in the particular portion which he may choose to abstract from the stream and take into his possession, and that during the time of his possession only."

If, then, a Local Sanitary Authority, or generally the promoters of a water-supply scheme, desire to utilize any spring or stream of water for that purpose, they must first come to terms with the owner of the land whereon the spring rises, or the riparian owner of that part of the stream from which they wish to take their supply. As such owner can only grant the limited powers which he himself possesses, terms must then be arranged with all the riparian owners lower down the stream who have appropriated, or may appropriate, the water to a beneficial use. Such arrangements must be made in writing.

"If any works are proposed to be done affecting any water or water rights, the proper course to be observed by a Local Authority is to serve a notice under the 328th section of the Public Health Act, 1875, on all persons interested, specifying the particulars of the matters and things intended to be done. Then will follow, either the consent of the parties interested, or there will be a reference to arbitration, and then will follow, either compensation for injury, if the injury can be compensated in money, or an abandonment of the proposed works." There is recourse to the promotion of a private Bill in Parliament, but this is too expensive, except in the case of very large schemes.

Water rights require to be very carefully dealt with, and every detail should be settled before the works are commenced.

In the first instance, the quantity of water should be gauged, by methods to be described later, the fall of the stream ascertained, and a careful investigation made of the use to which the water is put below the point of the proposed intake. The quantity of water required by any mills or other machinery worked by the stream should be obtained, as well as the quantity used for domestic, dairy, and other purposes.

The head waters of a stream are usually required for the purposes of waterworks when storage is necessary. It therefore becomes a matter for serious consideration whether such abstraction, either wholly or partially, will deprive the riparian owners below the site of the proposed works of sufficient water for the purposes of fishing, irrigation, mills, factories, boundary fences, and other matters. The compensation in such cases is either given in kind or in money. The quantity in the former case varies from one-third to one-tenth of the available annual yield of the gathering ground, necessitating in numerous cases the construction of " compensation reservoirs." The object of these reservoirs is to store the flood water, so as to maintain a continuous or intermittent flow, as may be arranged or settled by Provisional Order or Act of Parliament. When the proprietors are disposed to treat, and are not numerous, it is better in most cases to purchase their rights and be relieved of a large proportion of the compensation water.

There is no property in underground water, and any proprietor may sink or dig wells and obtain water, even if by doing so the water in a neighbouring proprietor's well is abstracted or diverted into another channel. In some cases wells are sunk to the spring supplying a stream, which by intercepting the spring at a higher point may considerably reduce the flow of water in the stream.

"But although a landowner will not in general be restrained from drawing off the subterranean waters in the adjoining land, yet he will be restrained, if, in so doing, he drains off the water flowing *in a defined surface* channel through the adjoining land."

As an instance of the application of this principle, the following case is given. A large provincial water company recently promoted a Bill to enable them to sink wells at various points along a stream, and pump the water which percolated into them to a storage reservoir. The riparian owners and residents in the valley opposed the Bill with such success that the quantity of water to be allowed to flow in the stream before pumping could proceed was so great as to render the Act when passed useless to the water company.

Much of the above information has been extracted from the valuable legal works of Messrs. Michael and Will, and Messrs. W. C. & A. Glen, to whom thanks are due.

EASEMENTS AND COMPENSATION.

In laying the various mains and branches in connection with a system of water supply, it frequently becomes necessary to lay pipes or tunnel through private property. In cases where the purchase of the freehold of the land is not made obligatory or desirable, an easement is obtained.

The term "easement" is defined in an ancient work called the "Terms de la Ley" as follows : "An easement is a privilege that one neighbour hath of another by writing or prescription, without profit, as a way or sink through his land or such like."

An easement giving the right to lay pipes, build culverts, drive tunnels, etc., for the purpose of conveying water, implies a right of entry at all times for repairs or other purposes rendered necessary for its proper enjoyment. And, further, the person to whom the easement is granted may prevent the owner of the land from doing anything to interfere with such a right—as, for instance, building houses or planting trees over the line of easement, or otherwise placing any obstruction to the full and proper enjoyment by the purchaser thereof.

In the case of Pomfret *v.* Ricroft, Twysden, J., observed as follows : "If a man gives me a licence to lay pipes of lead in his land to convey water to my cistern, I may afterwards enter and dig the land to mend the pipes, though the soil belongs to another and not to me. Whoever grants a thing is supposed also tacitly to grant that without which the grant itself would be of no effect."

In the case of Goodheart *v.* Hyett, the owner of the land commenced building over the line of easement, and the owner of the easement sought to restrain him from doing so, on the ground that if the house was built it would be impossible, or not reasonably practicable, for the owner of the easement to have access to the pipe for repairs. The Court restrained the owner of the land from building over the line of easement.

In taking into consideration the compensation due to the owner of the land, the elements of claim for damage may be classified as follows :—

1. The privilege of carrying a certain thing, either continuously or intermittently, over or under the land to the profit of the purchaser.

2. The right of entry to pipe or culvert at any time.

3. The right of preventing any buildings or other

obstructions from being erected or built on the line of easement.

4. The interference with the profitable laying out of the land for building sites or otherwise.

5. The driving of the tunnel or excavation of the trench may have the effect of withdrawing the moisture from the crops, and thus depreciate the value of the land for agricultural purposes. On the other hand, it may be of great benefit where the land is marshy or waterlogged.

6. The driving of the tunnel or excavation of the trench may intercept or divert underground water which previously had risen on other portions of the land.

The methods adopted by experts, and the results arrived at, differ so materially as to render it impossible to give any common data upon which the valuation is based. This is chiefly due to the variety of opinions held as to what constitute the elements of damage, even on adjoining land when precisely the same conditions prevail.

The following are, however, a few of the methods adopted by experienced valuers :—

1. The length of the easement in yards is multiplied successively by 33 feet, the value per acre, and thirty years' purchase.

2. The length of the easement in yards is multiplied successively by 4 yards, the value per acre, and from forty to fifty years' purchase, and the result divided equally between the two parties to the easement.

3. The length of the easement in yards is multiplied by 8 yards, and then by half the value of the fee simple.

To the results arrived at by any of the above methods a fixed price must be added for ventilating shafts, air-valve standards, or other surface arrangements, varying from 5s. to £6 each.

The widths taken for easement for the purposes of valuation are somewhat elastic, varying from 6 feet to 66 feet. For pipes or culverts up to 2 feet diameter, the width is frequently taken as 12 feet ; for larger culverts the width taken generally varies from 12 feet to 33 feet.

The average price is taken by some valuers at £3 per chain lineal, irrespective of width. Other valuers take from 3*d.* to 2*s.* per lineal yard for agricultural land, and from 5*s.* to £6 per lineal yard for building land, as the average price throughout the length of the easement.

A Local Authority has full powers as to laying pipes, etc., along the highways under its control within its own district. With regard to highways outside its district, should any persons having the care of such highways object in writing, pipes must not be laid without the consent of the Local Government Board. The easement is usually granted conditionally upon the undertakers reinstating the road and keeping it in repair for one year after the completion of the work, to the satisfaction of the authority granting the easement. In some cases the authority agree to accept a fixed sum per mile, relieving the undertakers from any further liability.

CHAPTER V.

GRAVITATION.

It has already been stated that a gravitation supply should always be adopted where possible, especially for rural districts.

The principal requirements of a gravitation scheme are as follows :—

1. That the spring or source of supply is situated at a sufficient elevation with regard to the place to be supplied, so as to produce a velocity in the pipes sufficient to deliver the quantity of water required.

2. That the intervening ground along the proposed line of pipes, between the source of supply and the district to be supplied, does not rise appreciably above the hydraulic mean gradient of the system.

3. That the pipes are selected of such dimensions as will discharge the requisite quantity without necessitating a greater velocity than 3 feet per second.

4. That sufficient storage-room is afforded, so as to allow for exceptional demands upon the supply, as well as for diminution in the latter in very dry seasons.

The subject of the flow of water in pipes has been so elaborately dealt with in the various text-books, that only the leading principles affecting actual practice will be dealt with here.

In the annexed figure (Fig. 1) a pipe is shown connecting two reservoirs A and B, in each of which the water is always kept at the same level. Vertical pipes, C_1, C_2, C_3, C_4, open at the upper end, are attached at intervals to the pipe AB. If the extreme end B of the pipe AB be closed, the water will stand in the pipes C_1, C_2, C_3, C_4, at the level of the horizontal line AD, through the surface of the water in the reservoir A.

As soon, however, as the end of the pipe at B is opened, and the water is allowed to flow uninterruptedly from the reservoir A to the reservoir B, the level of the water in the pipes C_1, C_2, C_3, C_4, will sink to E_1, E_2, E_3, E_4, respectively. The line connecting the points E_1, E_2, E_3, E_4 is called the hydraulic mean gradient, or virtual slope of the system, and if the pipe AB be of uniform section throughout its length, the hydraulic mean gradient will be a straight line joining the surface of the water in A and B.

This gradient is represented by the height of the reservoir A above the reservoir B, divided by the length of the pipe, which is the sine of the angle made by the line of the gradient with the horizontal. It has been found that the velocity acquired by water flowing through a pipe varies directly as the square root of the quantity representing the hydraulic mean gradient, and directly as the square root of the diameter of the pipe.

The section of the lines C_1, C_2, C_3, C_4, situated between the line of pipe AB, and the hydraulic mean gradient denotes the pressure (in addition to the atmospheric pressure) in the pipe at those points, and it is evident that when the line of pipe and the hydraulic mean gradient coincide, the pipe may be replaced by an open channel.

Suppose, however, that the line of pipe rises above the hydraulic mean gradient, it is clear that the pressure at that point is less than the atmospheric pressure by an amount indicated by the distance of the pipe above the hydraulic mean gradient. When this distance exceeds the height of a column of water which can be supported by the atmospheric pressure (34 feet), the pressure becomes nil, and flow ceases. Practically, the distance should never exceed 25 feet. When the pressure at any point F (Fig. 2) is less than the atmospheric pressure, the flow continues by syphonage until sufficient air is extracted from the water, which fills the summit of the pipe, and syphonage ceases. The pipe at this point may then be replaced by an open channel. The pipe is practically divided into two sections, AF, FB, and

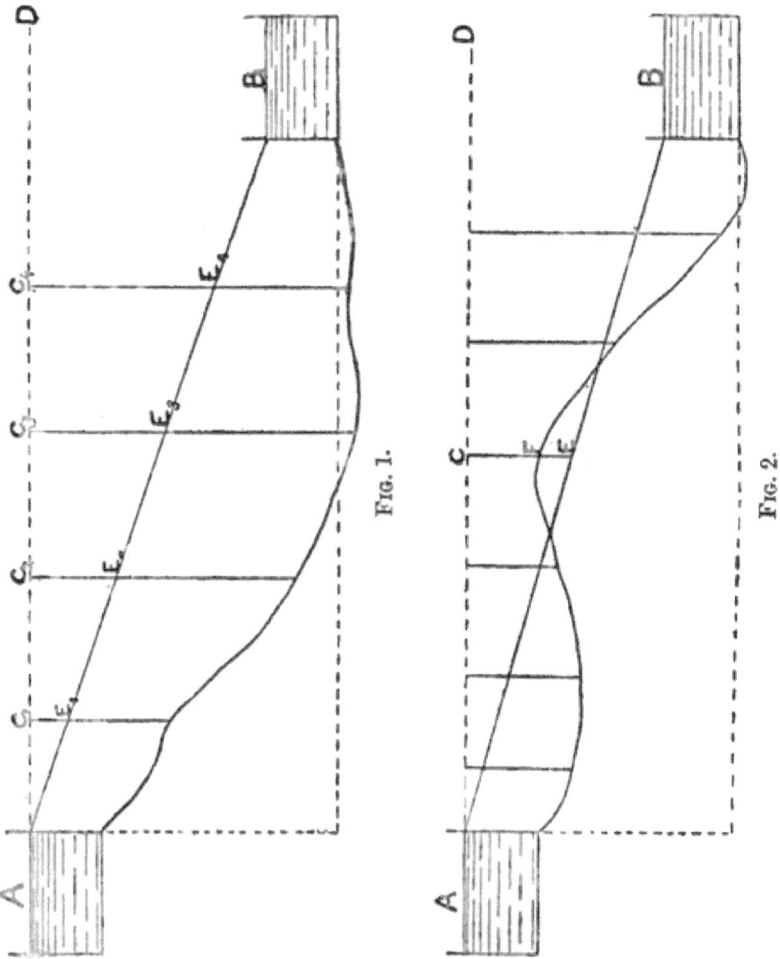

FIG. 1.

FIG. 2.

the discharge at B must depend upon the quantity which the first section AF is capable of delivering at the point F.

As the velocity, and, therefore, the discharge, depend upon the diameter of the pipe and upon the hydraulic mean gradient, a deficiency in the latter may be made up by an increase in the former. In the case just cited, the diameter of the pipe between A and F should be increased, so that, with the gradient AF, the same volume of water may be delivered at the point F as the portion of the pipe between F and B is capable of discharging with the gradient FB.

When the end of the pipe B (Fig. 1) was closed, the level of the water in the pipes C_1, C_2, C_3, C_4 rose to the same horizontal level as the water in the reservoir A. The hydraulic mean gradient will, therefore, rise or fall between the lines AD, AB, according as the orifice at B is closed or open.

In practice the orifice at B is never constantly open to its full extent, or, in other words, the maximum quantity that the main is capable of discharging is not constantly being discharged. Advantage is taken of this fact in the use of air-valves, which are fixed at the summit of all sections of the main which rise above the hydraulic mean gradient (calculated upon a maximum discharge). Whenever the consumption is less than the maximum, the hydraulic mean gradient rises, forcing the air out of the summits of the system and allowing syphonage to recommence, when the hydraulic mean gradient falls back to its lower position.

The formulæ generally used for calculation of velocity and discharge of water in pipes are :—

1.
$$v = 39 \sqrt{di}$$

Where v = velocity in feet per second.

   ,, d = diameter of pipe in feet.

   ,, i = height of point of supply above point of discharge (h) divided by the length of the pipe (l) ($= \frac{h}{l}$), each of which must be referred to the same unit—say feet.

2.
$$g = \sqrt{\frac{(3d)^5 \times h}{l}}$$

Where g = discharge in gallons per minute.

   ,, d = diameter of pipe in inches.

   ,, h = head in feet.

   ,, l = length of pipe in yards.

These formulæ allow a sufficient margin for subsequent rusting in the pipes.

By substituting the value 3 for v in the first equation, the following result is obtained—

$$d = \cdot 006 \times \frac{l}{h}$$

which enables the diameter of a pipe to be calculated where the head and length are known, so that the velocity may be three feet per second.

A gravitation supply usually requires a larger storage capacity than a pumping supply, as the sources are generally small at such elevations as will allow of gravitation. The amount of storage is regulated by the volume and permanency of the source. This matter, together with the subject of collection, will be dealt with subsequently.

When the source of supply is situated below the level of the immediately surrounding ground, and considerable expense would be entailed by excavation to a sufficient depth to allow a fall in the direction of supply, recourse may be had to a syphon. The summit of the syphon must theoretically not exceed 34 feet above the level of the surface of the water from which the supply is to be drawn ; practically about 25 feet is the limit. Such systems of supply have been adopted at Abingdon and Warwick.

The syphon may be charged by an air-pump attached to the longer leg, the communication with the main being cut off by means of a sluice-valve ; or the extremities of both legs may be closed by means of sluice-valves and the apparatus filled with water through a cock at its summit; the cock is then closed and the sluice-valves opened, when flow immediately commences.

When the water is highly aërated the syphon has to be frequently recharged. In calculating discharge, the head must be measured to the surface of the water in the reservoir, and not to the summit of the syphon.

CHAPTER VI.

PUMPING BY STEAM, GAS, PETROLEUM, WATER, AND WIND POWER.

THE various systems of motive power for pumping are determined by the work to be performed, the accessibility of the pumping station, and by local and other conditions depending upon the particular circumstances of each case.

The unit of power in common use is the mechanical force necessary to perform a certain amount of work known as a horse power, and is equal to 33,000 lbs. raised one foot high per minute. The terms used to express horse-power being somewhat indefinite, a brief reference may not be out of place.

Nominal horse-power is a commercial term for stating the size of an engine without regard to the actual power it will develop.

Actual or indicated horse-power is the power calculated from a diagram of the work performed by the steam in the cylinder, one horse-power being equal to 33,000 lbs. lifted one foot high in one minute, or—

$$\text{I. h.p.} = \frac{\text{Units of work done per minute}}{33,000}$$

Effective or brake horse-power is measured by a friction brake or dynamometer, and represents the actual horse-power less the power absorbed by the working parts of the engine or motor.

It is only proposed to refer to the indicated horse-power except where otherwise stated.

Steam power as applied to pumping for waterworks, is of general application, and the results are more economical for heavy pumping than any other system. The engine for applying the power takes several well-known forms, among which may be mentioned the vertical, horizontal, beam, and Cornish engines, each of which has several types. Steam-engines are divided into two main systems—non-condensing and condensing. The former exhaust their steam direct into the atmosphere, and the steam is used at full pressure, either partially or throughout the stroke, sufficient allowance being made to cut off and avoid back pressure. Condensing engines exhaust their steam into a chamber termed a condenser, which is in a state of partial vacuum owing to the steam coming in contact with a number of tubes through or around which cold water is circulating, or, in some cases, a jet of cold water. The air and condensed water are removed by an air-pump, which is worked from the engine. The water from the condensers having an average temperature of 100 degrees Fahrenheit, is frequently used for feeding the boilers, but care must be taken to prevent grease getting into it and injuring the boilers. Condensing engines are divided into systems according to the number of expansions employed —viz. simple or single cylinder, compound or two cylinders, triple or three cylinders, and so on as the range of expansion increases. The simple engine consists of a single cylinder in which the steam is exhausted by the condenser after having done its work. The compound engine consists of two cylinders; the steam after being partially expanded in the small or high pressure cylinder is exhausted into the large, or low pressure cylinder, and there undergoes further expansion before being exhausted by the condenser. The chief difference between the simple and compound systems is that in the former case the whole range of the tempera-ture occurs in one cylinder, whereas in the latter it is divided between the two cylinders, and the loss due to the extreme variation of temperature in one cylinder is thereby prevented. Theoretically the low-pressure cylinder with steam pressure and expansion the same as the high-

D

pressure cylinder worked on the simple or single system would develop more power than the two combined; but practically, owing to the various losses that occur, the theoretical results cannot be attained.

The advantages of the non-condensing engine are—

1. The simplicity of the mechanism and construction.
2. The easy accessibility to its working parts, and
3. Inexpensive foundations.

In fuel economy, however, it does not compare favourably with the condensing engine for permanent work. It is chiefly used in waterworks for temporary purposes, or where only a small engine is required.

The advantage of the condensing engine is its economy of fuel. The first cost is high, and the foundations are expensive, but for heavy pumping the satisfactory working of these engines, together with a fuel economy of about 25 per cent. over the non-condensing engines, outweigh any other considerations.

The Consumption of Coal per I.H.P. per Hour.

Non-condensing engines from 4 to 7 lbs.
Condensing engines (simple) from 3 to 5 lbs.
Condensing engines (compound) from 1½ to 3 lbs.

Gas power is utilized by the explosion of a mixture of coal gas and air in the cylinder, which, acting on the piston, gives the requisite motion. The charge consists of air next the piston combining gradually with a mixture of gas and air, which becomes stronger until the firing point is reached.

This gradual increase of explosive strength has the effect of doing the work gradually and preventing shocks, as well as sustaining the pressure at the end of the stroke.

Pumping by gas has many advantages over the use of small steam engines—

1. There is no loss when the engine is not working.
2. It can be started by merely turning the gas on and lighting the jet, at the same time giving the fly-wheel a start.
3. It can be fixed in almost any position, and requires no

attention, as must be the case when a boiler and steam engine are used.

There are many forms of these engines, each claiming special advantages, and all giving satisfactory results.

The consumption of gas per indicated horse-power varies from $17\frac{1}{2}$ cubic feet per hour in the larger engines to 25 cubic feet per hour in the smaller sizes.

Where the ordinary illuminating gas is either too costly or not available, the Dowson Gas Producers are frequently adopted, giving a non-illuminating gas which costs from $2\frac{1}{4}d.$ to $4d.$ per 1,000 cubic feet.

Petroleum-power engines differ from the gas engines chiefly in the method of delivering the oil in measured quantities with the requisite quantity of air. The oil is stored in a tank of sufficient capacity to serve for 12 or 24 hours as required. The firing light is obtained from the flame of a lamp kept continually burning. The advantages of this engine for pumping are—

1. The cheapness of the oil.
2. The slight amount of attention required.
3. The small capital cost.
4. The facility of fixing in any position.

The cost for oil varies from $\frac{3}{4}d.$ to $1\frac{1}{2}d.$ per indicated horse-power per hour.

Water-power may be utilized for pumping in several different ways, among which are hydraulic rams, water-wheels, and turbines.

The hydraulic ram is frequently applied when the water is abundant and the fall moderate. The action is as follows: The momentum of the inflowing water when arrested is expended in forcing a portion of itself through the delivering-pipe into a tank or reservoir.

If H = Height of source of supply above the ram.
  ,,   h = Height to which the water is to be forced.
  ,,   Q = Volume of supply.
  ,,   D = Volume delivered.

$$D = \frac{4Q}{7} \cdot \frac{H}{h}$$

The advantages of the ram are—
1. The simplicity of its parts.
2. The facility with which it can be fixed.
3. The little or no attention required.
4. Its moderate cost.

Water-wheels for driving pumps and other purposes are named according to the way in which they are acted on by the water.

1. *Overshot* when the water is delivered on the top of the wheel.

2. *Breast* when delivered about the centre, and—

3. *Undershot* when driven from the bottom, where there is a considerable velocity in the water.

The overshot wheel gives the greatest power, with the least expenditure of water, and is therefore applicable where the supply of water is scanty.

The horse-power (effective) is calculated as follows:—

$$\text{E.H.P.} = \frac{Q \times H \text{ in feet}}{C}$$

Q = Quantity of water in cubic feet per second.
H = Effective height of the fall in feet.
C = 13 for overshot wheels;
　　15 for breast wheels;
　　11·7 for high-breast wheels;
　　22 for undershot wheels.

Turbines, when carefully designed with regard to the conditions of working, are the best and most efficient motors. They are divided into two classes, pressure and impulse turbines, the former acting partly by impulse and partly by pressure, and the latter entirely by impulse. The turbine consist of a cylinder revolving horizontally, to which are attached spiral discs. The water is introduced at the top, and by its pressure on the sides and bottoms of the spiral chambers, causes the cylinder to rotate. The power is applied to the pumps by means of suitable gearing. In some cases an efficiency of 78 per cent. of the total power expended has been attained.

Actual horse-power $= \cdot 079 \, Qh$

Q = quantity of water passing through in cubic feet per second.

h = height of the fall in feet.

The theoretical horse-power contained in the water is calculated as follows :—

$$\text{T.H.P.} = \cdot 001892 \, Qh$$
$$Q = \frac{528 \cdot 5 \; \text{T.H.P.}}{h}$$

Q = quantity of water in cubic feet per minute.

h = head of water in feet.

EFFECTIVE HORSE-POWER FOR DIFFERENT MOTORS.

Theoretical power being	...		=	1·00
Turbine	...	...	=	·70
Overshot wheel	...	...	=	·68
High breast wheel	...	...	=	·60
Hydraulic ram	...	...	=	·60
Breast wheel	...	...	=	·55
Undershot wheel	...	...	=	·35

Wind-power is only economical for intermittent work, or where sufficient storage is provided for two or three days' supply. The wind pressure may generally be depended upon for seven or eight hours per day. The modern windmills for pumping are self-adjusting, and give exceedingly good results. They are being largely adopted for private supplies, or where their economical use permits.

$$\text{H.P.} = \frac{A \; V^3}{1,100,000}$$

A = Total area of sails in square feet.

V = Velocity of the wind in feet per second.

Table of the efficiency of windmills working eight hours per day, with a wind velocity of fifteen miles per hour during pumping.

Diam. of mill.	Revolutions per minute.	A.H.P. developed.	Quantity raised to a height of 100 feet.
Feet.			Galls.
12	55	$\frac{1}{4}$	3,375
15	50	$\frac{1}{3}$	5,000
18	45	$\frac{1}{2}$	10,000
20	40	$\frac{3}{4}$	12,500

TABLE OF WIND VELOCITY.

Velocity in feet per second	12·13	17·15	21·	24·25
Velocity in miles per hour	8·27	11·69	14·31	16·53
Description of Wind ...	Gentle	Slight breeze	Fresh breeze	Strong breeze

CHAPTER VII

VARIOUS FORMS OF PUMPS.

THE force acquired by steam or other motive power may be applied through the medium of a pump, in three ways: firstly, by suction or lifting; secondly, by forcing; thirdly, by a combination of the two systems, lifting and forcing.

The suction or lift pump (Fig. 3) is of common application for domestic supplies from wells or boreholes. It consists essentially of a cylinder or working barrel, with a suction pipe at the lower end, at the top of which is a valve, technically called a "clack." The delivery-pipe, or rising main, is attached to the upper end of the barrel, and through it a pump-rod, with a valve or bucket attached to its lower end, is worked up and down in the working-barrel. The upward movement of the bucket withdraws the pressure of the atmosphere from the surface of the water inside the suction pipe, and the pressure of the atmosphere on the surface of the water in the well forces the water up above the clack, or to such a height that the pressures on either side of the pipe are in equilibrium.

The water retained by the clack passed through the valve at the lower end of the pump-rod as the latter moves downwards, and is raised at each successive stroke until it reaches the top of the rising main or delivery-pipe, which in open-topped pumps is the top pump-tree or pipe. In small pumps, where the pump-rod works through a stuffing-box or gland (Fig. 4), the water can be raised to any required height, but for economical purposes it is not advantageous to lift it higher than 30 feet above the top of the pump. The height of the clack in the upper end of the suction-pipe should not exceed 25 feet above the lowest level of the water in the well, the best results being obtained from 10 feet to 15 feet, and in high speeds the shorter the suction the greater the efficiency. Although theoretically, when the barometer is standing at 30 inches, the water should rise in the suction-pipe to a height of 33·99 feet from the surface of the water (the specific gravity of mercury being $13·596 \times 30 \div 12 = 33·99$), it is impossible in

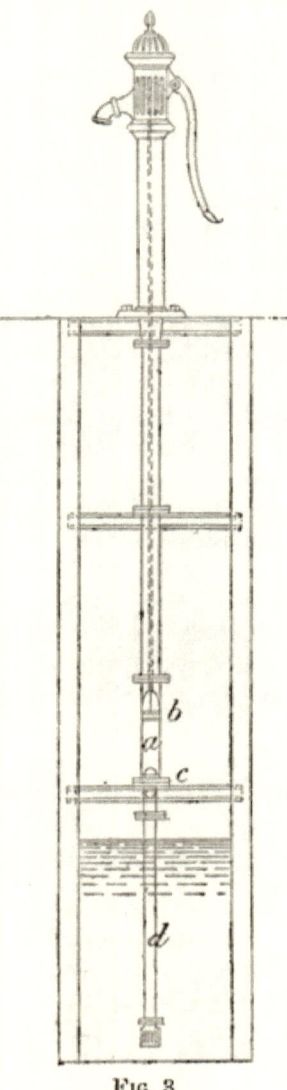

Fig. 3.

a, Working barrel; *b*, bucket; *c*, clack; *d*, suction.

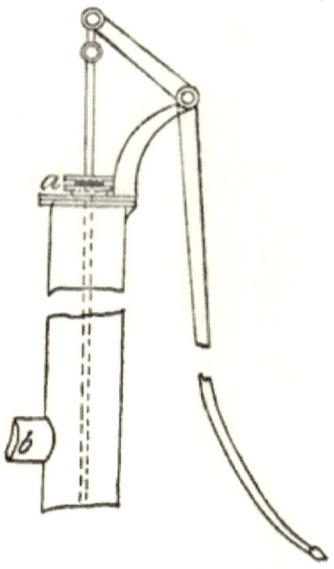

Fig. 4.

a, Stuffing-box; *b*, delivery-main.

practice to obtain so perfect a vacuum as to allow the water
to rise this height, owing to the variations of atmospheric
pressure, imperfect joints, and the friction of the pump.
The power to work the lift-pump is transmitted by rods
working from a beam, or by bell-cranks, or, in a few cases,
by direct action, as in the Bull engine. When worked by
manual labour, either a lever or wheel and handle are used.
In deep wells the lift-pump is generally used to lift the
water to a tank at the surface, from which it is taken by
force, or bucket and plunger pumps, and delivered at the
height required. The price of the ordinary lift-pump, with
a working barrel, 3 inches in diameter, complete, for a depth
of 30 feet, is £5, and from 2s. 6d. to 2s. 9d. per foot beyond
that depth. The capacity of this pump when worked by
hand is equal to 400 gallons per hour, lifted from a depth
of 30 feet. A double or "two-throw" pump (Fig. 5), the
diameter of the barrels being 3 inches and the stroke 10
inches, worked from the surface by rods, and driven by a
horse and gearing, will cost about £35, including 20 feet of
suction pipe and 50 feet of rising main (or delivery-pipe),
and air-vessel complete.

Approximate quantities in gallons raised per hour by
single, double, and treble-barrel pumps working at a
uniform speed of 20 strokes per minute—

Dim. of pumps.	Length of stroke.	Single barrel.	Double barrel.	Treble barrel.
Inches.	Inches.			
2½	9	165	330	495
3	9	240	480	720
3½	9	310	620	930
4	10	480	960	1440
4	12	575	1150	1725
5	12	900	1800	2700
5	15	1125	2250	3375
6	12	1280	2560	3840
6	15	1600	3200	4800
6	18	1920	3840	5760

These quantities raised assume the horse to travel at

an average rate of three miles per hour, the pumps thus making twenty strokes per minute with the single speed gear. Pumps up to 4 inch barrel may be worked at 30 strokes per minute; unless the height the water has to be

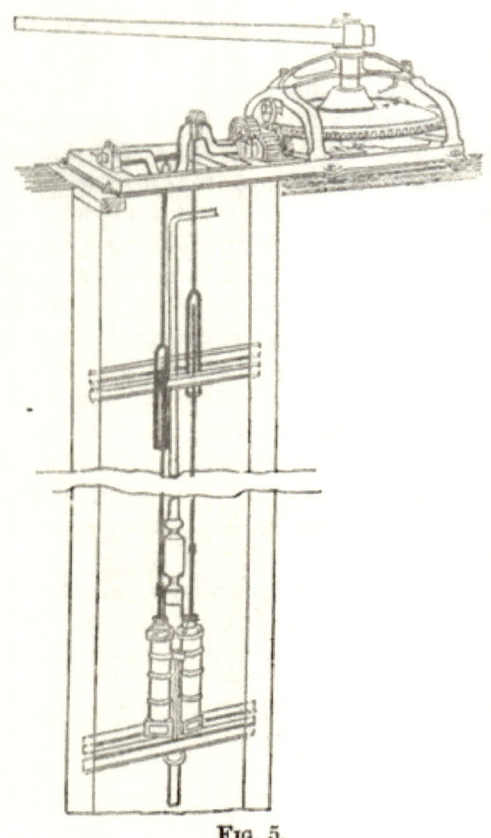

Fig. 5.

raised is great, the speed of the larger sizes should vary from 20 to 25 strokes per minute according to the lift.

The plunger or force-pump (Fig. 6) consists of a cylinder or working barrel, in which the piston or plunger works up and down through a stuffing-box or gland. The plunger is either hollow or of solid metal, according to the conditions required, and may consist of one or more plungers, each working in its own barrel. The working barrel is of cast iron (or preferably of brass or bronze), and connected at one end with the delivery pipe, with valve-box and air-vessel beyond. The suction-pipe and valve-box are at the other end. This pump works either horizontally or vertically, and its action is as follows: in the up-stroke of the plunger a vacuum is created which allows the water to enter through the suction-pipe into the working barrel and body of the pump, filling the space left by the plunger. The water is retained by a clack or valve at the top of the suction-pipe, and is again forced by the downstroke of the plunger through the delivery-pipe, being retained by the delivery-valve, and rises at each successive stroke until it reaches the point of discharge. During the up-stroke of the plunger

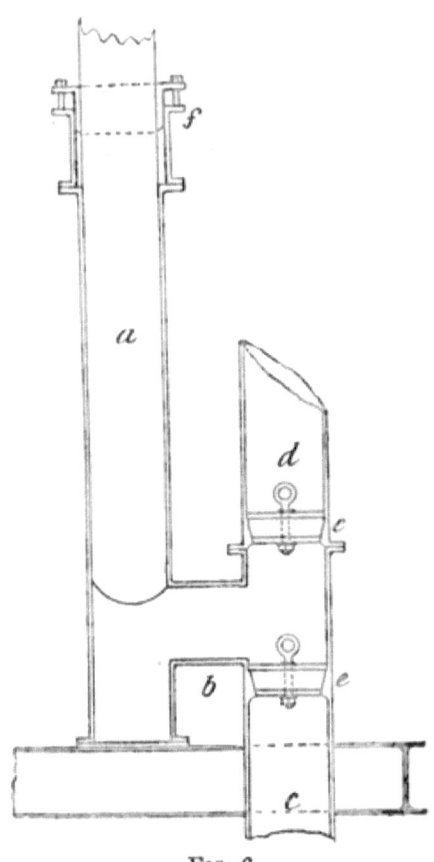

Fig. 6.

a, Plunger; *b*, H-piece with doors; *c*, suction; *d*, rising main; *e*, valves; *f*, stuffing-box.

the forward motion of the water through the delivery-pipe would cease, and the discharge would therefore become intermittent instead of continuous. This is avoided by the

use of an air-vessel, the air of which being compressed
during the descent of the piston or plunger re-acts and
forces the water through the delivery main during the up-
stroke.   This prevents the shock to the working parts
caused by the force which would be required to overcome
the inertia of the water, and at the same time economizes
the power of the engine by keeping the water in constant
motion.

The plunger or force-pump possesses great advantages
over the lift-pump in most cases where it can be employed,
and is especially suitable where considerable height has to
be overcome, or where continuous working is required.   It
is not suitable in positions where the water is likely to rise
above the pump, owing to the difficulty of access to the
working parts in case of accident.   In deep wells, therefore,
the lift-pump with open top is to be recommended for the
deep-pumping, and the plunger for subsequently raising the
water to the required elevation.

The lifting and forcing pump, or bucket and plunger combined (Fig. 7), was invented by Perkins, and introduced at
the Lambeth Waterworks in 1848.

The construction is similar to that of the forcing-pump
described above, except that the ram or piston has a bucket
attached by a rod to its lower end.   The upper portion is
enlarged to form the ram, having a sectional area equal to
one-half that of the working barrel.   The theoretical quantity of water which rises into the working barrel at each
up-stroke of the bucket is equal to the capacity of the barrel
through which it ascends, one-half of which quantity rises in
the delivery or rising main on the descent of the bucket, and
the remaining portion is discharged during the following
up-stroke.   The delivery from the pump is therefore continuous.   This is one of the best forms of double-acting
pump, as it possesses nearly all the simplicity of the single-
acting pump, and is free from the defects of the four-valve
pumps.   The quantity of water delivered at the up and
down stroke is no more than with a pump with single
action, the difference being that the double-action gives a

continuous, and the single-action an intermittent delivery. This form of pump is used in nearly all the large waterworks.

Horizontal engines and pumps, mainly direct acting, are frequently used for small supplies, and improvements during the last few years have justified their use in many of the largest waterworks. The Worthington, Deane, Davidson, and other well-known types give good results; they are principally used for forcing to service reservoirs.

The pump-trees or pipes which constitute the suction-pipe and rising main, form so important and costly a portion of the pumping apparatus as to require careful design. They are usually 9 feet long, and consist of cast-iron pipes with flanges, or of wrought-iron or steel tubes, riveted or welded with flange-joints. The cast-iron pipes should be made of hard mottled-grey iron, re-melted in the cupola, and cast vertically in loam, care being taken to keep the metal of uniform thickness and truly cylindrical. In open-topped

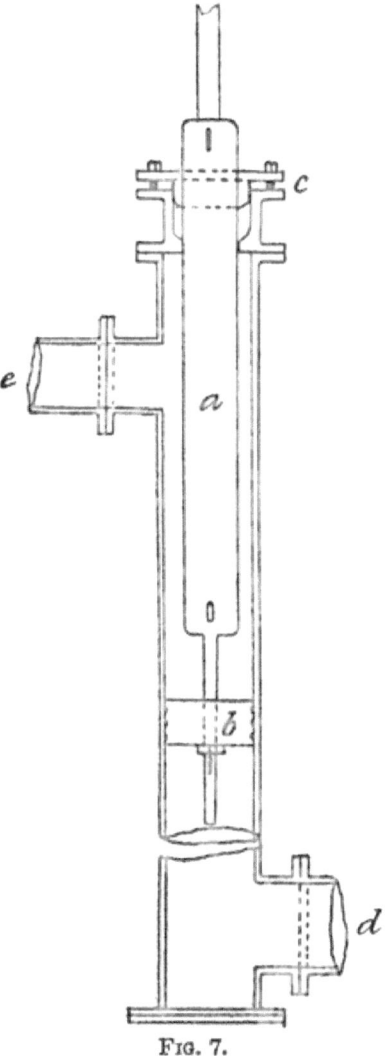

FIG. 7.

a, Plunger; b, bucket; c, stuffing-box; d, connection to valve chamber; e, rising main with stop back valve and air vessel.

pumps the diameter is made from ¼ inch to 1½ inch greater than the working barrel, so as to enable the valves to be withdrawn when repairs are necessary. In close-topped pumps working through a stuffing-box or gland the rising main is usually two-thirds the area of the working barrel. The pipes, whether constructed of cast or wrought iron, or of steel, have the flanges machine-faced so as to be perfectly plumb when bolted together. The joints are either made with red lead or flannel steeped in tallow, and bolted tightly together. In considerable heights of rising main, it is the usual practice to reduce the thickness of metal every 54 feet. The wrought-iron or steel tubes for large diameters should be riveted together with a butt-strap joint, and the flange formed with an angle-iron shrunk on the body of the pipe and riveted. The smaller sizes are usually welded or solid-drawn. The pipes should all be painted, or, in the case of small tubes, galvanized.

The following table gives the weight and thickness of cast and wrought-iron pipes:—

Cast-Iron flanged Pipes.				Wrought-iron Tubes.	
Diameter.	Thickness.	Weight of 9 feet lengths.		Thickness.	Weight per foot.
Inches.	Inches.	cwts.	lbs.	B. W. G.	lbs.
1½	—	—	—	14	1·37
2	—	—	—	14	1·81
2¼	—	—	—	12	2·24
3	⅜	1	12	11	2·68
3½	—	—	—	10	3·11
4	⅜	1	49	10	3·55
4½	—	—	—	10	3·98
5	7/16	2	10	9	4·42
5½	—	—	—	9	4·85
6	7/16	2	65	9	5·29
7	½	3	32	—	—
8	9/16	3	81	—	—
9	9/16	4	80	—	—
10	9/16	5	23	—	—
11	9/16	5	79	—	—
12	⅝	6	103	—	—

CHAPTER VIII.

VARIOUS FORMS OF PUMPS—*continued*.

THE working-barrel is formed of hard grey metal, bored out truly cylindrical in the larger pumps, and with a gun-metal or copper liner inserted in the smaller ones. The ends are slightly bell-mouthed, and are made sufficiently long to allow from 3 inches to 12 inches clearance beyond the actual stroke of the pump. The thickness of the metal is greater than that of the rising main, owing to the wear and tear, and to allow for reboring when necessary. In forcing-pumps the top of the barrel is made tight, with a stuffing-box or gland packed with metallic material, which can be renewed without stopping the working of the pump.

The clack or waist-piece contains a turned conical seating for the valve. In some cases, in open-topped pumps, a second seating of larger diameter is provided above the one generally in use. The advantage of this arrangement is that it enables a temporary valve to be lowered, in the event of an accident, to act until access can be had to the defective valve. Door-pieces are fixed so as to enable the bucket or clack to be examined or changed. These, of course, are only available when the water is below the level at which they are situated. The diameter of the suction-pipe may be reduced below the level of the clack or valve to from one-half to two-thirds of the area of the working-barrel, except in the case of quick-running pumps, when the diameter should not be less than that of the working-barrel.

The rose, windbore, or strainer, at the bottom of the suction-pipes takes various shapes; but care must be taken to make the aggregate area of the apertures not less than from two to two and a half times the area of the suction-pipe.

The pump-rods are either made of wrought-iron with flanged joints bolted together, or of pitch-pine connected by means of iron side-plates and bolts running through the rods. Hard-wood guides are affixed to the rods when working in the rising main, and metal rollers guide the rods when working a plunger.

The valves, of which there are at least two in every pump, either fixed or movable, require the most careful attention, as they frequently cause a large portion of the power of the pumping apparatus to be lost. It is essential that they should offer little resistance to the passage of the water in one direction, and close the passage quickly and entirely in the contrary direction, so as to prevent slip. The weight of the valve should be sufficient to close without knocking, and be light enough to be lifted without offering undue resistance to the water. In high lifts the valve is usually calculated at 1 lb. in weight per square inch of area, equal to 2·3 feet of water; and for low lifts it varies from ¼ lb. to ½ lb. per square inch of area. The velocity of the water through the valves should not exceed 5 feet per second. The valves used in pumps belong to one of two classes, the hinged or door, and the spindle valve.

The flap or hinged valve (Fig. 8) consists of a flap or

FIG. 8.

sheet of leather, stiffened and weighted with metal plates, working on a hinge, the shell being of wood or metal.

The butterfly valve (Fig. 9) is of frequent application,

and derives its name from the wings or flaps, consisting of semi-circular discs hinged to the centre of the shell. The

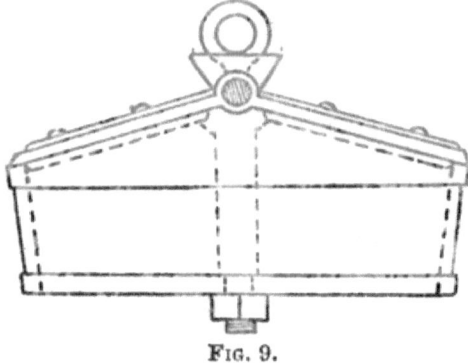

Fig. 9.

wings, or flaps, are stiffened and weighted with metal plates, similar to the flap, or hinged valve, and the shell is formed of wood or metal.

The mitre valve (Fig. 10), used mainly in horizontal

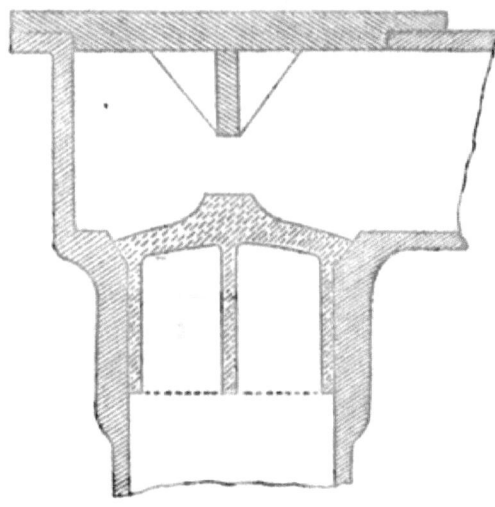

Fig. 10.

pumps, consists of a circular metallic disc, with conical

E

face, the upper portion having a short spindle to limit its
lift, and feathers below to guide the valve on to its seat.

The rubber disc valves (Fig. 11), both single and double,
are largely used for lift-pumps, and consist of an iron or
gun-metal seat or grid, either forming part of the shell or
fitting into a recess in it. The rubber forms the valve, and
is prevented from rising too high by the guard shown in
the figure. The apertures in the seat or grid are placed
at an angle to produce a circular motion in the water and
thence in the valve. This prevents the valve from falling

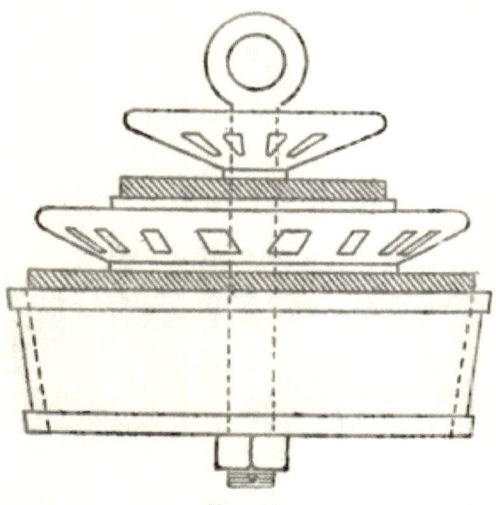

FIG. 11.

in the same position and gradually cutting the rubber.
Sufficient clearance must be allowed around the spindle for
the lift of the valve. Strong dark blue rubber, which is
a little heavier than pure rubber, stands better for heavy
work.

The double-beat valve (Fig. 12) was first introduced by
the well-known firm of Harvey & West, of Hayle, for the
Wicksteed Cornish engine at the East London Waterworks.
It was designed to overcome the battering and the great
wear and tear of the flap-valves, and has been used with

little modification up to the present time. It consists of a circular ring, on which the lower part of the valve beats, and a similar ring of less diameter on the plate of which the upper part beats, forming "the double beat." The

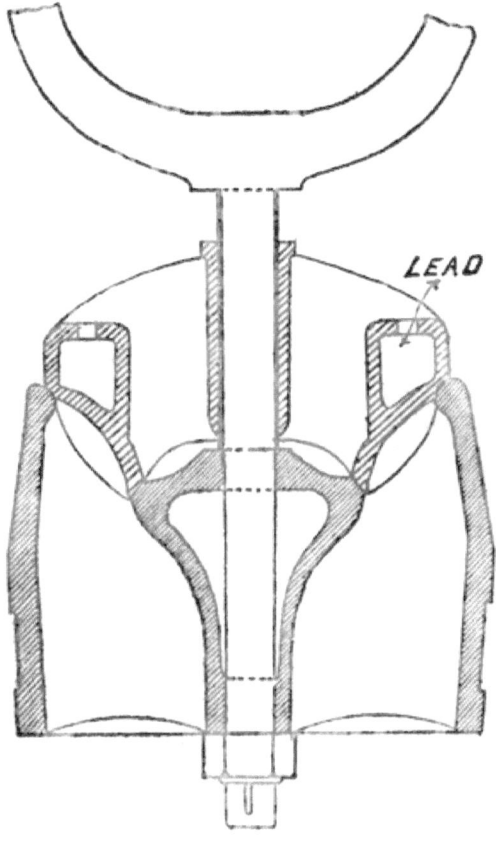

FIG. 12.

beats are formed of lignum-vitæ, white metal, or leather. They are fixed to the valve and beat on a gun-metal seat. The valve consists of a double cylinder, one within the other, forming one piece, open top and bottom, and working on a spindle.

The webs connecting the parts together are placed at a slight angle, to cause the valve to rotate during the influx of water every time it rises and falls, which keeps the beats perfect and tends to prevent grooving. The great advantage of this valve is its small lift, owing to the two openings for discharge, the vibration caused by the closing of the valve being diminished in consequence.

The three- and four-beat valves only present slight variations, due to the multiple number of beats.

The Riedler valve, which is being largely used on the Continent for quick-speed pumps, is giving great satisfaction. Professor Riedler, the inventor of the valve, has made a series of observations on the action of valves (Indicator-Versuche au Pumpen), and the valve referred to is the result of his investigations, with a view to remedying the defects of valves used in quick-speed pumps. The lift is performed automatically, and the closing of the valve is accomplished by a spring and lever.

The air-vessel is simply a cylindrical vessel of cast- or wrought-iron, with a dome-top and inlet and outlet connections from the pump at or near the base. The air in the upper part of the vessel is compressed until it balances the pressure of the water being pumped. At those parts of the stroke at which the motion of the pump piston exceeds the average speed, the surplus water further compresses the air to a small extent, and is thereby received into the air-vessel; again, at those parts of the stroke where the speed of the piston is below the average, the water thus stored up in the air-vessel is forced out by the expansion of the air, and supplies the deficiency. The air-vessel equalizes the strain on the pumps and pipes through which the water flows, and renders the delivery nearly constant, and is to the flow of water what the fly-wheel is to an engine. The only trouble in practice is that of keeping the air-vessel charged with air, as compressed air in contact with water is more or less rapidly absorbed. Provision should be made either by having a small pump for the purpose of pumping air into the air-vessel, or a small cock should be fixed on the suction pipe to allow air to be pumped with the water.

The quantity of water contained in a pipe is determined by the formula $x = \cdot00283d^2l$, in which x = quantity in gallons, d = diameter of pipe in inches, and l = length of pipe in inches; being based on the fact that the area of a circle in square inches is $\cdot7854d^2$, and a gallon contains 277 cubic inches, therefore $\dfrac{\cdot7854d^2l}{277} = \cdot00283d^2l$; and $x \times 10$ = weight in lbs., the weight of a gallon of water being 10 lbs.

To find the pressure of water in a pipe.—The pressure in lbs. per square inch is determined by the formula, $x = \cdot433h$, in which x = the pressure in lbs. per square inch, h = the head or height of water in feet; being based on the fact that a cubic foot of water weighs 62·4 lbs., and a square foot contains 144 square inches, therefore $\dfrac{62\cdot4}{144} = \cdot433$ lbs.

The quantity of water delivered at each stroke of a pump is obtained by means of the formula given for the contents of a pipe. If d = the diameter of the working barrel or plunger in inches, l = the length of stroke in inches.

The quantity delivered per minute is determined by the number of strokes per minute multiplied by $\cdot00283d^2l$.

These calculations make no allowance for slip, which varies from 5 to 15 per cent. or more, according to the condition of the pumps. The amount of slip is determined by accurately measuring the quantity of water delivered by the pumps into a cistern or tank of known capacity, or over a weir, and comparing the quantity with the calculated delivery, according to the dimensions of the pump. The method of gauging over weirs will be described in a future chapter.

CHAPTER IX

RAINFALL, SPRINGS, STREAMS, AND THEIR MODE
OF MEASUREMENT.

The moisture which is constantly being evaporated from the
sea and other water surfaces is carried by the air, and is
returned as rain and snow to feed the springs and streams,
and is again, in turn, evaporated to supply the sources of
rainfall. There is always moisture present in the air,
varying according to the season and situation. In this
country the average proportion is stated by eminent authorities to be about 1½ per cent. When there is a considerable
amount of moisture in the air, approaching saturation, a
slight reduction of temperature causes the moisture to become visible in the form of mist, rain, hail, or snow.

The rainfall of a district depends largely upon the position
of its mountain ranges and forests, together with the direction
of the prevalent winds. It is a matter of common observation in this country that the western and southern shores
have an annual rainfall considerably in excess of the eastern
shore. The rainfall on the western coast varies from 40
to 70 inches per annum, and in an exceptional case, in 1883,
the great depth of 190·28 inches was recorded at the Stye, in
Cumberland. The rainfall on the southern coast varies
from 30 to 40 inches, and on the eastern coast from 20 to 30
inches, with an extremely low rainfall in 1883 of 18·71
inches at Clacton, in Essex.

The distribution of rainfall is very variable over the
surface of the globe, due to the peculiar conditions

prevailing in each district. In Great Britain, the average fall is about 33 inches, with considerable variation. Spain has about 100 inches on the Atlantic Coast, and from 8 to 10 inches at Madrid; India from 10 to 600 inches; in Australia the average fall is about 25 inches; North America, 40 to 90 inches; South America, from *nil* up to 270 inches; and in Africa, from *nil* up to 40 inches. The areas over which an occasional shower falls at long intervals, exceeding a year in many cases, and termed rainless districts, are the deserts of North Africa, Arabia, Persia, Beloochistan, Thibet, Mexico, Guatemala, California, Peru, and Chili.

There are many cases recorded of excessive rainfall occurring within short periods in this country, causing the bursting of embankments, and floods of great magnitude, where sufficient means of discharging the flood-waters, at such times, is absent. Buchan states that 3 inches is not infrequently recorded in 24 hours in the Highlands of Scotland. At Seathwaite, in Cumberland, 6·62 inches fell on November 27, 1848. A remarkable fall was recorded at the Newport Waterworks reservoir during the 24 hours ending 9 a.m. on July 15, 1875, of 5·33 inches; the same storm was the cause of two small reservoir disasters, viz. at the Rogers Pond reservoir, at Cwm Carn, Monmouthshire, and at the Blakeney Brook reservoir at Cinderford, in Gloucestershire.

In estimating the available rainfall for water supply, it is the minimum rainfall on which all the calculations must be based. Mr. G. J. Symonds, F.R.S., who has done such valuable service to the country in organizing the complete system of record of rainfall observations, gives the following proportions as the limits of fluctuation in the rainfall. These are the result of a large number of observations extending over many years, and which, he states, will be within 7 per cent. of the actual fall :—

Wettest year, 45 per cent. more than the average.

Driest year, 33 per cent. less than the average.

Driest two consecutive years, 26 per cent. less than the average.

Driest three consecutive years, 21 per cent. less than the average.

The rainfall of a district is estimated from actual measurement by means of rain-gauges. These observations should be taken daily at or about the same time—usually 9 a.m. The number of guages for any district depends upon the extent of the catchment area or water-shed, and upon its altitude. The "Snowdon" pattern of rain-gauge (Fig. 13) is frequently adopted for waterworks purposes. It has a

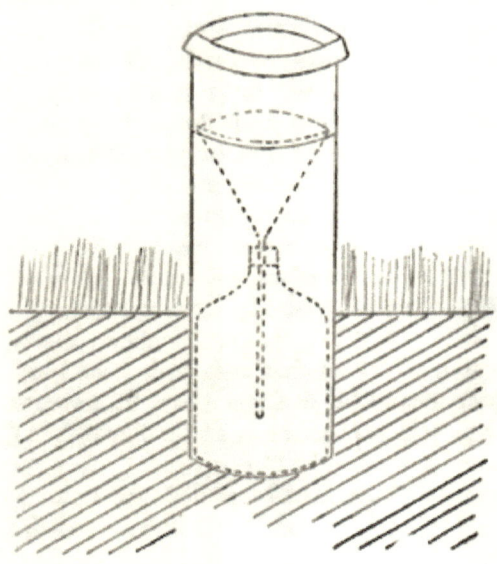

Fig. 13.

diameter of 5 or 8 inches. It should be placed on a level base of stone or similar material, and either a recess cut in the stone to admit the gauge, or pegs driven around the cylinder to prevent lateral movement. The site should be on even ground, with plenty of open space, and at a distance of not less than one and a half times the height measured horizontally from plantations or buildings. The top of the gauge should stand 12 inches clear of the ground. The gauge consists of a copper cylinder, provided with a funnel of the same metal to receive the rain and prevent evaporation. The tube of the funnel terminates in a glass cylinder, which retains the rain-water until the observer has measured the depth of water in a graduated glass measure for the purpose. The altitude of the rain-gauge should be carefully taken by connecting its height with the nearest ordnance bench-mark.

The rain and snow which fall on the surface of the earth are disposed of in the following ways:—

1. Evaporation.
2. Percolation.
3. The remainder flows off as storm or flood-water.

1. *Evaporation*, or the property of the water to rise in vapour, has received the attention of many eminent observers, among whom may be mentioned the names of Evans, at Nash Mills; Gilbert and Lawes, at Rothampstead; and Greaves, at Lea Bridge, near London. The latter found the annual evaporation from large water surfaces was about 21 inches, and was distributed in the following proportion during the year:—

January to March	...	...	...	4 inches.
April to May ...	...	...	...	8 ,,
June to September	...	...	...	7 ,,
October to December ...	...	...	2 ,,

The amount of evaporation from large water areas is in most cases equal to the rainfall, therefore in estimating the yield of a water-shed or catchment, the area of the lake or reservoir should be excluded. The evaporation from land surfaces varies according to the geological and physical conditions prevailing in the district, and any fixed rule is impossible. Steep slopes in the lower series of rocks afford the greatest flow over the surface, and little evaporation and percolation. Plantations lessen the loss by evaporation. The amount of evaporation on land surfaces in this country varies from 8 to 20 inches, and in hotter climates is much greater; in parts of India it varies from 55 to 90 inches, and on water surfaces the average is usually taken at 72 inches.

The Dead Sea and the Mediterranean afford examples of great evaporation.

2. *Percolation*, or the passage of rainfall through the surface of the ground, varies also according to the geological and physical conditions. The first recorded experiments in this country were made towards the end of the last century by Dr. Dalton, of Manchester, and have been continued by Dickenson and Evans, Gilbert and Lawes, Greaves, Latham, and others. In the evidence before the Royal Commission on Metropolitan Water Supply, 1893, the following particulars were given as to the percolation through 3 feet of soil, with grass growing on the surface, and 3 feet of chalk, also with grass growing on the surface, at Nash Mills, Hemel-Hempstead, and to these are appended the results obtained from Lea Bridge and Rothampstead :—

District.		Period.	Medium.	Rainfall.	Percolation.
			Feet.	Inches.	Inches.
Nash Mills ...	...	1842 to 1884	Soil 3	27·40	6·77
,, ,, ...	...	1854 to 1884	Chalk 3	27·84	10·55
Lea Bridge ...	...	1852 to 1873	Soil 3	25·94	7·02
Rothampstead	...	1871 to 1892	Soil 5	30·11	13·90

In the Rothampstead experiments a solid block of earth was enclosed in a water-tight tank, and in each case the experiments were made with level surfaces.

3. *The storm or flood water* varies according to the absorbent power of the ground over which it flows, together with the amount of evaporation. It either flows off the surface to form streams and rivers, which supply many towns, such as London, York, and Chester, or it may be impounded in the head waters for the supply of towns at a distance. The average summer flow of water-sheds with rocks of medium absorbing power and steep slopes does not, as a rule, exceed 3·12 gallons per 1000 acres per second. In times of flood the flow off such water-sheds requires special precautions. Heavy rainfalls, causing excessive floods, have occurred

during the construction of reservoir dams, and are not by any means unusual : among others may be mentioned the following :—

Reservoir.	Owner.	Date.	Galls. per second per 1000 acres.	Rainfall per 24 hours (at the rate of).
				Inches.
Woodhead ...	Manchester Corporation	Oct., 1849	3125	12·
Rhodes Wood	,, ,,	Feb., 1852	1562	5·96
Vyrnwy ...	Liverpool Corporation	Jan., 1883	1112	4·24
Vartry ...	Dublin Corporation ...	—	3202	12·22
Tansa ...	Bombay Corporation ...	—	4640	17·71

An inch of rainfall per 24 hours per 1000 acres is equivalent to 42·01 cubic feet, or 261,951 galls. per second.

1 in. per acre = 100 tons = 22,400 galls.

Although such extraordinary floods are of short duration, and occur at intervals of some years, yet the circumstances attending them must be taken into consideration in the design and construction of reservoir works, by providing means of passing such floods through the works without endangering them, and that such means of exit should be at all times clear, without the aid of manual or mechanical labour being required.

The average daily flow of some of the large rivers is given below—

		Per day.
River Thames at Ditton	906	million galls.
,, Severn	300	,, ,,
,, Ouse, at York ...	140	,, ,,
,, Tiber (Italy) ...	5500	,, ,,

To arrive at an accurate estimate of the quantity of water available in a catchment area, it is necessary to have rain-gauges fixed as previously stated, and recorded every day with simultaneous gaugings of the flow of water in the streams and springs. Careful attention must be given to the stratification and dip of the rocks, as it is by no means an infrequent occurrence for a large portion of the rainfall to follow the dip of the strata and rise as springs in an adjoining water-shed. The gauging of the rainfall, streams, and springs, should extend over as long a period as possible, in order that the necessary calculations may be based on reliable data. The experiments and results given as to evaporation and percolation are instructive and interesting from a scientific point of view, but have been carried out on too limited a scale to be relied upon for the general purposes of water engineering.

CHAPTER X.

MEASUREMENT AND ESTIMATION OF THE FLOW OF WATER.

THE units of measurement usually adopted in gauging the flow of water, are the cubic foot and gallon for capacity, and a minute, second, or twenty-four hours for time.

An imperial gallon of water at a temperature of 62° Fahr. and a barometric pressure of 30 inches, weighs 10 lbs; and a cubic foot contains 6·235 (practically 6¼) gallons.

The flow of water through sluices, pipes, or channels, is governed by the same laws as falling bodies, and its motion would be uniformly accelerated but for the resistance offered by the friction and form of the channel.

The theoretical velocity due to the force of gravitation, friction being neglected, is expressed by the formula—

$$v = \sqrt{2gh}.$$

Where v = velocity in feet per second.

    ,, g = the force of gravitation, or the velocity acquired by a body falling through space under the influence of the attraction of the earth, in one second.

    ,, h = the head, vertical distance through which the water has fallen, or difference in level of the two ends of the channel, in feet.

The numerical value of g varies slightly according to the altitude and the latitude. In England the value usually adopted is 32·2 feet per second. The above formula may therefore be written—

$$v = 8{\cdot}025 \sqrt{h}.$$

If to the natural head artificial pressure equivalent to h′ feet has been added, then—

$$v = 8{\cdot}025 \sqrt{h + h'}.$$

These formulæ require modification according to the particular form of orifice through which water is discharged.

Gauging by means of an Orifice.

The discharge of water through an orifice is proportional to the area of the orifice and the mean velocity of discharge. Theoretically the discharge from an orifice should be equal to the product of the velocity of discharge and the area of the orifice. Experiment has shown, however, that the converging currents of water as they approach the aperture, produce a contraction in the area of the issuing stream, varying in degree according to the form of the orifice. This is called the *vena contracta.* A coefficient, determined by experiment, has therefore to be applied in each case, so as to make allowance for this contraction.

The formula for discharge through an orifice may therefore be written—

$$q = 8{\cdot}025 \, c \, a \, \sqrt{h}.$$

Where q = discharge in cubic feet per second.
,, a = area of orifice in square feet.
,, h = head in feet, or the height of the surface of the water above the centre of the orifice.
,, c = a coefficient applicable to the particular form of orifice.

The following values for c are adapted from those given in Spon's Engineering Tables :—

Round or square orifices in a thin plate, ·62.

Sluice at end of a rectangular channel, ·70.

Short tubes (three diameters and under) with square edges, ·81.

Short tubes when the tube projects into a reservoir or cistern, ·71.

The following table give the results of experiments made

by eminent observers upon circular orifices, with sharp inner edges :—

Name.				Head.	Diameter of orifice.	Coefficient.
				Feet.	Inches.	
Abbé Bossut	...	...	...	0·6	1·0	0·649
Castel	...	...	...	2·7	1·2	0·629
Venturi	...	...	...	2·9	1·6	0·622
Rennie	...	...	...	1·0	1·0	0·633
Rennie	...	...	...	2·0	1·0	0·619
Eytelwein	...	...	...	2·4	1·0	0·618
Weisbach	...	...	...	2·0	1·2	0·614

Mr. Mair Rumley, in his experiments at Messrs. Simpson and Co.'s works at Pimlico, recorded in the "Proceedings" of the Institution of Civil Engineers, found that the coefficient of discharge was affected by the temperature of the water.

The discharge through a submerged orifice is calculated in exactly the same manner, except that the difference in level of the surface of the water on either side of the orifice is taken as the head.

Gauging by means of Weirs.

For this purpose a sharp-edged weir (Figs. 14 and 15) gives the most satisfactory results. A still-water pond should be formed on the up-stream side of the weir, to steady the flow of the water. A peg should be driven at a point in this pond as far as possible from the weir, and the upper surface of the peg should be made perfectly level with the upper edge of the weir. As it is difficult to drive a peg with precision under water, especially when the bottom of the pond is hard and stony, the following is a useful practice : Drive the peg so that its upper surface is slightly below the required level, and then drive a long flat-headed nail into the top of the peg. By means of a hammer the nail may be easily driven until its head is exactly level with the upper surface, or sill, of the weir.

To construct a weir for the purpose of ascertaining the discharge of a stream of water, a water-tight dam must be formed, the best material for which is clay. In this dam the weir is fixed, which usually consists of a plank or frame

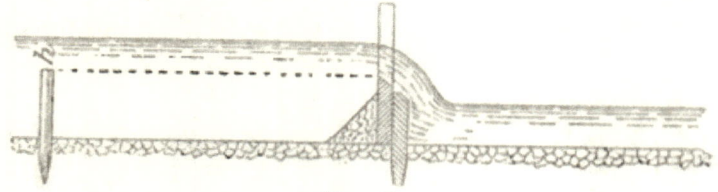

FIG. 14.

of wood, with a rectangular notch cut in its upper edge. The plank is kept in a vertical position by means of stakes driven on either side of it. The horizontal edge of the notch over which the water flows, must be fixed perfectly level, and must be bevelled so as to present a thin edge on the up-stream side. The depth of the water below the

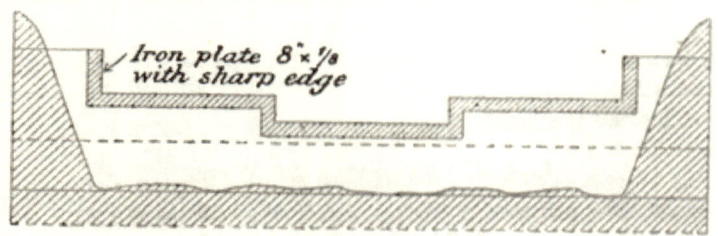

FIG. 15.

sill of the weir on the up-stream side should not be less than three times the depth of the water flowing over the weir; and the difference in level between the surface of the water on the down-stream side of the weir and the sill of the weir should not be less than half the maximum depth of the water flowing over the weir.

The theoretical formula for the discharge of water through rectangular notches is—

$$Q = \tfrac{2}{3} l \sqrt{(2\,g)}\ \mathrm{H}^{\tfrac{3}{2}}$$

Where l = length of notch in feet.

H = height in feet of the free-level of the discharging water above the sill.

Q = discharge over weir in cubic feet per second.

Owing to the interference with the free flow of the stream occasioned by the ends and sill of the notch, a coefficient, c, has to be applied to this equation, bringing it to the form—

$$Q = \tfrac{2}{3} cl \sqrt{(2g)} H^{\frac{3}{2}}$$

The coefficient c varies with l and H.

With values of H between ·25 and 2, and with l not less than 2, the coefficient c is fairly constant, and may be taken as ·62, which is the same as that for the discharge of water through round or square orifices in a thin plate, given above.

In cases where extreme accuracy is not required, the following formula, proposed by the late Mr. Thomas Hawksley, F.R.S., may be employed :—

$$Q' = \frac{lh \sqrt{h}}{2}$$

Where Q' = discharge over weir in gallons per second.

„ h = depth of water flowing over weir in inches.

„ l = length of notch in feet.

The table given on p. 66 has been calculated from this formula.

All measurements of depth should be taken at the peg above referred to, which should be situated at least 3 feet above the weir. A thin steel rule should be used for this purpose.

Where, however, only the approximate discharge is required, the measurement may be taken over the sill of the weir. This method will obviously give a low discharge.

Gauging by means of Uniform Channels.

The calculation of the discharge by uniform channels, such as canals and bye-washes, is of great importance in waterworks engineering, and has received much attention.

At the commencement of this chapter it was stated that the flow of water through sluice-pipes or channels is governed by the same laws as falling bodies, and its motion would be uniformly accelerated but for the resistance offered

F

DISCHARGE IN GALLONS PER TWENTY-FOUR HOURS FOR EACH FOOT IN WIDTH OF SILL.

Head of Water. Inches.	Decimals of an inch.										Head of Water. Inches.
	.0	.1	.2	.3	.4	.5	.6	.7	.8	.9	
0	—	1365	3863	7098	10,927	15,273	20,075	25,298	30,910	36,881	0
1	43,200	49,838	56,785	64,028	71,559	79,360	87,429	95,751	104,322	113,139	1
2	122,186	131,462	140,963	150,679	160,610	170,758	181,104	191,651	202,402	213,339	2
3	224,467	235,779	247,283	258,960	270,832	282,864	295,068	307,452	319,997	332,714	3
4	345,600	358,632	371,824	385,191	398,711	412,380	426,194	440,170	454,284	468,553	4
5	482,976	497,548	512,246	527,088	542,072	557,219	572,479	587,971	603,423	619,078	5
6	634,884	650,788	666,894	683,094	699,439	715,899	732,473	749,186	766,008	782,966	6

by the friction and form of the channel. The principal part of the friction is proportional to the square of the velocity, and is nearly the same at all depths. The friction, however, varies according to the surface of the fluid exposed to the solid in contact therewith, in proportion to the whole quantity of fluid; that is, the friction for any given quantity of water is as the surface of the bottom and sides of a river directly, and as the whole quantity of water in the river inversely. Therefore, supposing the whole quantity of water to be spread on a horizontal surface equal to the bottom and sides, the friction is inversely as the height at which the river would then stand, which is called the "hydraulic mean depth" (Eytelwein's "Hydraulics"). The hydraulic mean depth may be simply stated as the sectional area of a stream divided by its wetted perimeter.

Perhaps the most generally useful formula is that devised by Eytelwein, and slightly modified by Beardmore—

$$v = 55\sqrt{h \times 2f}$$

Where v = velocity in feet per minute.
  „ h = hydraulic mean depth, in feet.
  „ f = fall in feet per mile.

This formula must, however, be used with caution.

The following formula is given in Box's "Hydraulics" for long channels, neglecting head due to velocity of entry, which in long channels is inappreciable:—

$$C = \left(\frac{874520 \times F \times A}{L \times P} \right)^{\frac{1}{2}} \times A$$

Where L = length of channel in yards.
  „ A = cross-sectional area of stream in square feet.
  „ P = wetted perimeter.
  „ F = fall in inches.
  „ C = cubic feet discharged per minute.

GAUGING BY MEANS OF SURFACE VELOCITY.

The discharge of a stream may be found by observing the surface velocity by means of a wooden float or weighted tube. The time occupied by the float in passing over a measured distance (which should be as great as possible) is noted, and the velocity reduced to lineal feet per second. As the surface velocity in the centre of a stream is greater than the mean velocity of the whole body of water, a proper allowance must be made.

The proportion which the mean velocity of the water in a stream of tolerably uniform section bears to the surface velocity at the centre has been made the subject of much investigation. The following formulæ, amongst others, have been proposed :—

U = Mean velocity in feet per second.

V = Surface velocity at centre in feet per second.

1. Prony—

$$U = \frac{V\,(V + 7{\cdot}783)}{V + 10{\cdot}345}$$

2. Neville (for velocities less than 10 feet per second, in small channels)—

$$U = {\cdot}816\,V.$$

3. Boileau (depth not exceeding 1 foot)—

$$U = {\cdot}785\,V \text{ to } {\cdot}865\,V.$$

4. Beardmore—

$$U = (V + 2{\cdot}5) - \sqrt{5\,V}$$

The discharge is found by multiplying the mean sectional area by the mean velocity of the stream.

On a large scale an instrument called a current-meter is frequently used to determine the velocity, and hence the discharge of a stream.

The results arrived at by the above methods are only to be adopted when more reliable data cannot be obtained.

Very small streams may be gauged by allowing the water to flow into a vessel of known capacity (*e.g.* a *pail* or *cistern*), and noting the time taken in filling.

MEMORANDA.

Cubit feet per minute—

$$\times\ 9000 = \text{gallons per 24 hours.}$$
$$\text{Gallons} \times\ 1604 = \text{cubic feet.}$$
$$\text{Cubic feet} \times\ 6{\cdot}25 = \text{gallons.}$$

CHAPTER XI

PLANS, SECTIONS, LEVELLING, NECESSARY DATA.

HAVING investigated the available sources of suitable water in the neighbourhood of the district to be supplied, made careful gaugings, levellings, and obtained all possible information as to permanency of supply, compensation for water rights, probable demands for easements or purchase of land, and the nature of the ground with regard to excavation for reservoirs, laying mains, etc., it becomes necessary to embody or record the results obtained in the forms of plans, sections, and reports.

Water engineers have nowadays much to be thankful for in being able to obtain at a low cost the accurate surveys afforded by the Ordnance Department, instead of having to make special surveys for themselves—always tedious and often unnecessary. The Ordnance Surveys are issued on several scales, and can be obtained from Mr. Edward Stanford, Charing Cross, London, S.W., sole agent for England and Wales. The following information is extracted from a small pamphlet issued gratis by Mr. Stanford :—

1. $\frac{1}{500}$ (= 10·56 feet to a mile) for towns with population over 4000. Some towns have been published on the scales of $\frac{1}{528}$ (= 10 feet to a mile), and $\frac{1}{1056}$ (= 5 feet to a mile).

Each sheet represents 24 chains by 16 chains. Price, uncoloured, 2s. each ; coloured, 2s. 6d. to 10s. 6d.

2. $\frac{1}{2500}$ (= 25·344 inches to a mile). Each sheet represents $1\frac{1}{2}$ miles by 1 mile. Price, uncoloured, 2s. 6d. each (with areas printed on, 3s.) ; coloured, 2s. 6d. to 23s.

Approximately one square inch on these plans equals one acre.

The area of each enclosure, together with a reference number, is printed within it on the plan. The brace S on the plans indicates that the spaces so braced are included under the same reference number. Areas are computed to the centre of the fence or other boundary of the enclosure, except in the following cases :—

1. When the fence or other boundary is also the boundary of a parish or other civil division which does not follow the centre of the fence, the area is calculated to the parish or other boundary, and not to the centre.

2. The fences, etc., bounding either side of a railway are included wholly within its area.

Altitudes are given in feet above the approximate mean water at Liverpool. Those indicated thus ⬦ B. M. 54·7 refer to marks made on buildings, walls, etc., and are called bench-marks. Trinity high-water mark, which is the level of the lower edge of a stone fixed in the face of the river wall on the east side of the Hermitage entrance of the London Docks, is 12·48 feet above Ordnance datum.

3. Six inches to the mile. Each sheet represents six miles by four miles. For certain counties quarter sheets, which are reductions of the $\frac{1}{2500}$ plans, may be obtained; these represent three miles by two miles. Price, full sheets, 2s. and 2s. 6d., quarter sheets 1s. each.

4. One inch to the mile. Each sheet represents eighteen miles by twelve miles. Price (with one or two exceptions) 1s. each.

Before ordering plans an index map of the county, parish, or town in question should be obtained from Mr. Stanford. This will greatly facilitate the purchase, and save much delay and annoyance. It is generally best to have the sheets mounted on brown holland before they are sent. The charge is not heavy, and the results are excellent.

The $\frac{1}{2500}$th, commonly known as the 25 inches to the mile, scale is usually the most suitable for the general plan of a waterworks. Upon this plan the position of reservoirs and pumping-stations, and the lines of mains and branches are marked, the dimensions of the pipes being figured above them. The positions of sluice-valves, air-valves, hydrants, etc., are also indicated. The names of the owners and occupiers of all lands upon which it is proposed to construct works or lay pipes should be written in the enclosures; and the names of owners and occupiers of mills, or other property in connection with which claims may be made as regards riparian rights, should also be entered against the property in question.

Careful levellings must be made along the proposed lines of pipes, and these should be plotted to the same horizontal scale as the general plan and to a vertical scale of 20 feet to the inch.

Detail plans and sections of reservoirs, pumping-stations, etc., should be drawn to a scale of not less than eight feet to an inch.

The hydraulic mean gradients should be drawn upon the sections of the main and branch pipes.

The following example will show the method of calculation by which the losses of head due to friction, and hence the hydraulic mean gradients, are found. Suppose the storage reservoir to be situated at A (Fig. 16) and that 8640 gallons per day are to be delivered at the point C, 11,520 at the point E, and 14,400 at the point F. The lengths of the main and branches are shown on the section; also the levels at each point.

As the demand during the summer is frequently greater than that in the winter, and the demand during the middle of the day much exceeds that of the remainder, it is usual to take three times the average rate of supply as the basis upon which the diameters of the mains and branches are calculated. Reducing the rate per day to gallons per minute, the system must be so designed as to enable 18, 24, and 30 gallons per minute to be discharged at the points C, E, and F, respectively, with the head available.

Assume a 4-inch pipe from A to B.

Assume a 3-inch pipe from B to D.

Assume a 2-inch pipe from D to F.

Draw the horizontal line AA' through A and produce the ordinates through B, D, C, F, E, so as to cut AA' at B', D', C', F', E, respectively.

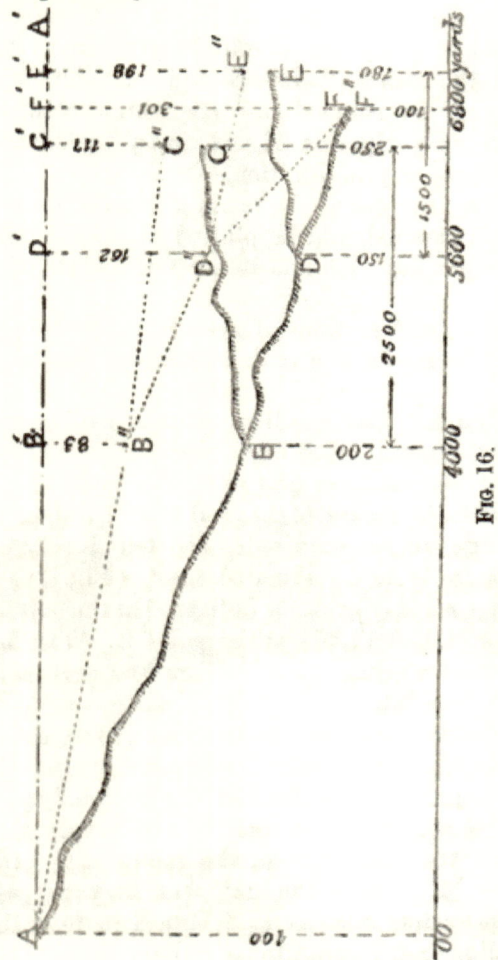

Fig. 16.

Then by the formula $G = \sqrt{\dfrac{(3d)^5 \times H}{L}}$ (or in practice by a set of tables), the loss of head due to the discharge of

—18 + 24 + 30 = 72 gallons per minute through the pipe AB is found to be 83 feet.

Make B'B" = 83 feet and join AB". Then AB" is the hydraulic mean gradient for the section AB, provided that no point in the pipe AB rises above the line AB".

In the same manner the loss of head due to the discharge of 24 + 30 = 54 gallons per minute through the pipe BD is 79 feet. Make D'D" = 83 + 79 = 162 feet, and join B"D", which is the hydraulic mean gradient for the pipe BD.

Again the loss of head due to the discharge of 30 gallons per minute through the pipe DF = 139 feet. Make F'F'" = 83 + 79 + 139 = 301 and join D"F'", which is the hydraulic mean gradient for the pipe DF.

The point F is therefore only 1 foot above the hydraulic mean gradient at that point, and this may be neglected as insignificant. It now remains to determine the diameters of the branch pipes, BC and DE.

The head at the point B, or the distance of the hydraulic mean gradient above that point, being 400 − 200 − 83 = 117 feet; and the point C being 250 − 200 = 50 feet above the point B; the available head at the point C is 117 − 50 = 67 feet. The diameter of a pipe 2500 yards in length to discharge 18 gallons per minute with a head of 67 feet is found from the same equation to be 2·184 inches.

A 2½-inch pipe (the commercial size next above 2·184) would therefore be used for this branch. The loss of head using a 2½-inch pipe, calculated in the same manner as in the previous cases, will be 34 feet. Make C'C" = 83 + 34 = 117 feet, and join B"C", which is the hydraulic mean gradient for the pipe BC.

The head at the point D = 400 − 150 − 83 − 79 = 88 feet; and as the point E is 180 − 150 = 30 feet above the point D, the available head at the point D = 88 − 30 = 58 feet.

Calculating, as in the last paragraph, the diameter necessary to discharge 24 gallons per minute at the point E = 2·277 inches, a 2½-inch pipe would therefore be used for this branch also, and the loss of head for such a pipe would be 36 feet.

Make E′ E″ = 162 + 36 = 198 feet, and join D″ E″, which is the hydraulic mean gradient for the pipe DE.

An examination of the section will now show that the pipes at no point rise above their respective hydraulic mean gradients, and that the latter are not situated unduly above the former.

The velocity of the water in these pipes necessary to obtain the specified discharges must now be calculated.

Let V = velocity in feet per second.

„ G = discharge in gallons per minute.

„ d = diameter of pipe in inches.

Then $V = \dfrac{G}{2\,d^2}$ (approximately).

From this formula the velocities may be calculated, and are tabulated as follows:—

Section.	Diameter of pipe in inches.	Discharge (gallons per minute).	Velocity (feet per second).
A to B	4	72	2
B to D	3	54	2
D to F	2	30	$3\frac{1}{2}$
B to C	$2\frac{1}{4}$	18	$1\frac{1}{2}$
D to E	$2\frac{1}{2}$	24	2

These velocities are below 3 feet per second (see p. 27), with the exception of the section D to F. It will be remembered that the point F was 1 foot below the hydraulic mean gradient. Taking both these circumstances into account, it would be advisable to increase the diameter of the pipe from D to F to $2\frac{1}{2}$ inches. This alteration will reduce the velocity in this section to $2\frac{1}{2}$ feet per second, and will raise the hydraulic mean gradient at F to 92 feet above it. The system may now be considered satisfactory.

Supposing that at any point the line of pipe had risen above its hydraulic mean gradient, then the pipes preceding that point would have to be enlarged to such an extent as would raise the hydraulic mean gradient above the pipe at that point. As an alternative, the line of pipe may be lowered by means of a tunnel or deep cutting, but this is rarely economical or expedient.

In calculating the diameters of the various pipes necessary to afford a stated supply at a given point, the head required at the point to be supplied must be taken into account. Supposing the level of the source of supply to be 500 feet above a certain village, but that a head of 100 feet is required at the village itself, to force the water to the top stories of the houses, or for fire-extinguishing purposes, then the available head upon which the diameter of the pipe leading from the source to the village must be based will be only $500 - 100 = 400$ feet.

In calculating the diameters of the pipes from the maximum daily rate of supply, a certain minimum must be observed. For instance, a branch may be laid to supply a block of three cottages containing a population of 15 people. Allowing 10 gallons per day for each person, the average rate of flow to afford this supply would be ·104 gallons per minute. Trebling this for a maximum daily supply would give ·312 (or less than $\frac{1}{3}$ gallon per minute), and the time taken in filling an ordinary 3 gallon bucket would be nearly ten minutes. A system should be arranged so that a discharge at a rate of 3 gallons per minute may be obtained at each connection. At farm-houses, where the water is used for refrigerating the milk, a supply of at least 3 gallons per minute is required.

On the other hand, it is frequently necessary to lay down a branch main of considerably larger diameter than would be necessary to afford the requisite supply. As this often leads to waste of water (especially where the place to be supplied is isolated), which may interfere with the hydraulic mean gradients of the entire system, reduced fittings may be enforced, or the supply may be regulated by a sluice valve under the sole control of the undertakers, fixed at the termination of the branch. The following formulæ will enable the engineer to calculate the amount of reduction necessary : —

Let g = discharge in gallons per minute.

   ,, l = length of branch in yards.

   ,, d = diameter of branch in inches.

   ,, d_1 = internal diameter of tap in inches.

   ,, h = total head in feet.

   ,, h_1 = head consumed by friction in branch.

   ,, h_2 = head consumed by a screw-down tap fixed at the termination of the branch.

Then $h = h_1 + h_2$ and let $d_1 = r\,d.$

Then assuming that the obstruction to the flow of the water caused by the tap is the same as in the case of short tubes (not less in length than twice the diameter of the orifice), the area at the point of greatest contraction will be ·81, the area of the passage through the tap.

Then (1) $h_1 = \dfrac{r^4 l}{1\cdot394d + r^4 l} \cdot h$

  ,, (2) $h_2 = \dfrac{1\cdot394d}{1\cdot394d + r^4 l} \cdot h$

  ,, (3) $g = \sqrt{\dfrac{(3d)^5 \times h}{l} \cdot \dfrac{r^4 l}{1\cdot394d + r^4 l}}$

As the calculations from which the diameters of the mains and branches of a waterworks are obtained are based upon the results of the preceding levelling operations, a few words upon levelling may not be out of place. In the first instance, the engineer should never neglect to test his level before commencing work. The testing need not extend further than to prove that the level will "reverse" truly, and that the line of sight lies in the plane of collimation.

If, when the instrument has been set up, the bubble does not remain stationary in the middle of the tube on the level being revolved, the error must be rectified half by raising or lowering, as the case may require, the level by means of the capstan-headed screws, and half by the parallel plate-screws.

If, however, the level, after the most careful adjustment, refuses to reverse correctly, it should be sent to the makers.

In the meantime, the bubble should be centred at each reading by means of the plate-screws.

To adjust the level for collimation, select a fairly level piece of ground and measure out three chains, and place pieces of flat stone or slate for the staff to rest upon at the commencement and terminations. Call these ABCD. Set up the level at B and read the staff on A and C. This will give the true difference of level between A and C, as the errors of adjustment would be the same on either side, the distance being equal, and would neutralize each other. Then set up the level at D, and read the staff again on A and C. If, on comparing the last reading with the previous ones, the difference of level between A and C is the same in each case, the level is in adjustment. If, however, the second operation does not agree with the first, an error is present. This error is proportional to the length of the sight, and is, therefore (in the second operation), three times as great on A as on C; half the error in level being the error due to one chain in distance. This error must be corrected by raising or lowering, as the case may require, the cross-hair of the level by means of the collimating screws, until the readings of the staff when placed on A and C give the same difference of level whether the instrument be set up at B or D.

Always read each sight twice—the second time after booking. In turning the level on its axis, always turn it the way of the sun, otherwise it may become unscrewed, which would cause error and probably serious delay.

The necessary data include the following :—

Yield and permanency of source.

Quantitative analysis of water.

Sanitary survey of source.

Lengths of mains and branches.

Population to be supplied.

Rate of supply per head.

Levels.

Cross sections and particulars of streams, etc., to be crossed.

Additional supplies for trade, dairy, or compensation purposes.

Names of owners and occupiers, and rateable value of land to be interfered with.

Nature of ground to be excavated.

Quality of local building materials and labour.

Prices of manual labour and horse hire.

Facilities for transit.

Particulars as to quantities of water used for mills, etc., in connection with which claims may be made.

CHAPTER XII

MATERIALS.

AMONGST the principal materials employed in waterworks construction the following are included: Iron (cast and wrought), steel, copper, lead, zinc, tin, brass, gun-metal or bronze, stone, bricks, concrete, cement, lime, gravel, sand, clay, and wood.

Cast and wrought iron and steel are made from "pig-iron," which is manufactured from iron ore. Pig-iron is the name given to the crude, unpurified metal in the form that it is first obtained from the blast furnaces, and is classified as: (1) Bessemer; (2) foundry; (3) forge.

Bessemer "pig" is dark-grey, contains a large proportion of free carbon, a small quantity of silicon and manganese, and is practically free from sulphur and phosphorus. It is principally used for conversion into steel (Bessemer process). Foundry "pig" contains a large proportion of free carbon, and is therefore specially adapted for foundry work. Forge "pig" contains little free carbon, and is therefore adapted for conversion into wrought iron.

Cast-iron is obtained by re-melting foundry "pig"-iron in a small furnace termed a cupola. Inferior castings are sometimes run direct from the blast furnace. Cast-iron is sub-divided into: (1) grey; (2) white; (3) mottled.

Grey cast-iron is made from the best foundry pig, and produces the best castings. White cast-iron is made from forge-pig, and is only used for the most inferior descriptions of castings. Mottled cast-iron is a mixture of the grey and white varieties. If a little nitric acid be applied to a clean fractured surface of cast-iron it will give a black stain with the grey variety and a brown stain with the white variety. White and mottled cast-iron do not rust so readily as the grey variety. Chilled castings are produced by using metal moulds.

Malleable cast-iron is produced by subsequently heating castings in an annealing oven with some substance containing an excess of oxygen. The oxygen combines with the carbon in the casting to a certain depth, depending upon the length of exposure, rendering that portion of the casting similar to wrought-iron.

Cast-iron contains a large percentage of carbon (from 2 per cent. to 6 per cent). It is chiefly used in waterworks construction in the manufacture of pipes, the bodies of sluice and other valves and appliances, in the heavy parts of engines and pumps, for roof-trusses, and for ironwork generally, which will not be subjected to tension.

Wrought-iron is generally prepared from forge-pig, by puddling, after which it is rolled and converted into " puddle-bars." The different qualities of " puddle-bars " are—

1. Puddled or rough bars.
2. Merchant bar, or common iron.
3. Best bar.
4. Best best bar.
5. Best best best bar.
6. Scrap bar; which is again subdivided into "best scrap," and "best best scrap."

Wrought-iron is practically free from carbon, and should not contain more than ·15 per cent. Wrought-iron is chiefly used in waterworks construction in the manufacture of tubes, for roof trusses, girders, and for ironwork exposed to tension, or where forging is necessary.

Steel is defined by Dr. Percy as " iron containing a small percentage of carbon, the alloy having the property of taking a temper." It contains from ·12 to 1·5 per cent. of carbon, and is, therefore, in composition, midway between cast and wrought-iron. Steel is manufactured in two ways—

1. By extracting a portion of its carbon from cast-iron (Sieman's process).

2. By adding carbon to wrought-iron (Bessemer process).

The principal varieties of steel are—

1. Blister steel.
2. Cast steel.
3. Mild steel.
4. Puddled steel.

Steel is chiefly used in waterworks construction for tubes, and largely in the manufacture of engines and pumps, for girders, and generally where high tensile strains have to be supported. The use of steel is rapidly superseding cast and wrought-iron for most engineering purposes.

Copper is principally employed for roses or strainers for pumps, or at outlets to reservoirs; also for floats for ball-taps, etc.

Lead is used for joining cast-iron pipes, for making distributing pipes, for weighting pump-valves.

Zinc is principally used for cisterns, covering roofs, and for covering iron to protect it from rusting (galvanizing).

All wrought-iron tubes used for distributing water should be galvanized. This is effected in the following manner. The iron is cleaned, and after being heated, is dipped in molten zinc, which forms a protecting coating without injuring the iron.

Tin is used for coating water-pipes internally, so as to protect them from the action of the water.

Brass is an alloy consisting of copper and zinc, in proportions varying from 2 to 18 of copper, to 1 of zinc. Brass is used for bushes and bearing surfaces, but is inferior to bronze. It is largely used for valves and taps.

Gun-metal or *Bronze* is an alloy of copper and tin.

Soft gun-metal contains 8 parts of tin to 92 of copper. Hard gun-metal, 18 parts of tin to 82 parts of copper. Bell-metal, $23\frac{1}{2}$, or 23, parts of tin, to $76\frac{1}{2}$, or 77, parts of copper.

Bronze is fusible, and makes good castings. It is soft, uniform in texture, and wears evenly, and is therefore specially suitable for bearing surfaces, producing little friction.

G

Its tenacity is high, and it does not corrode. It is extensively used for pump-barrels, valve-faces, slides, and screws, and also for making the best valves and taps.

The following table, taken from Unwin's "Machine Design," gives the safe limits of stress, with a live or varying load, to which most of the materials described above may be exposed :—

SAFE LIMIT IN POUNDS PER SQUARE INCH.

Material.	Tension.	Compression.	Shearing.
Cast-iron	3600	10,400	2700
Wrought-iron bars ...	10,400	10,400	7800
„ „ plates ...	10,000	10,000	7800
Soft steel, untempered	17,700	17,700	13,000
Cast steel „ ...	52,000	52,000	38,500
Copper	3600	3120	2300
Brass	3600	—	2700
Gun-metal (or bronze)...	3120	—	2400

Stone is used in waterworks construction, in building the walls of reservoirs and filter-beds, in the erection of engine and pumping houses, and as a constituent of concrete. Except, however, where stone is plentiful and easily worked, masonry is generally superseded by brickwork or concrete. Any hard stone may be used for concrete, though limestone perhaps gives the best results.

Bricks are almost indispensable in waterworks construction, for the purposes mentioned above in connection with stone, and often form a heavy item of expense.

The best quality of bricks should be used, especially for outside work. The cost of these is, however, frequently prohibitive, and then the local productions, if there are any, must be carefully inspected by the engineer. Economy only begins when efficiency has been attained; but the efficiency of a material depends upon what is required of it. The varieties of bricks depending upon the materials used and the subsequent manipulation are almost endless. For general work, however, the variety known as "stocks" will be suitable. These are hard-burned bricks, well-shaped and sound. Staffordshire blue-bricks are best for coping purposes. Good bricks should be sound, free from cracks and flaws, stones and lumps of any kind (especially lime). They should be regular in shape, uniform in size; their arrises or edges should be square, straight and sharply defined; their surfaces should be even, not hollow, and not too smooth. They should not absorb more than one-sixth of their weight of water. They should be hard, and burnt so thoroughly that there is incipient vitrifaction throughout. They should give out a ringing sound when struck against one another. The cost of brickwork is calculated by the rod or by the cubic yard. A rod of reduced brickwork consists of 272 superficial feet, one and a half bricks in thickness, and is equal to $11\frac{1}{3}$ cubic yards. The following estimate of the cost of brickwork for a small reservoir may be useful as indicating the points to be included.

COST OF BRICKWORK PER CUBIC YARD.

	s.	d.	s.	d.
400 stock bricks at 30s. per 1000	12	0		
Cartage	3	0		
			15	0
1 to 4 { 1½ bushel Portland cement at 3s. 9d. ...	5	0		
{ 5⅓ bushels sand at 1s. 1½d.	6	0		
			11	0
Labour—				
Bricklayer: 5 hours at 6d.	2	6		
Labourer: 5 hours at 3½d.	1	5½		
			3	11½
			£1 9	11½

Say 30s. per cubic yard.

In this instance the bricks were obtained from a local brickyard, and were of inferior quality, but good enough for the work. A thoroughly good class of bricks would have cost 66s. per thousand, including railway carriage and extra haulage. The Portland cement cost 10s. 2d. per cask, less 3s. for returned empties, or 43s. per ton nett, on rail in

London. There was no local sand suitable for mixing with cement. The sand estimated for was sea-sand, costing 4s. per ton on rail, and to this had to be added heavy railway carriage and haulage.

Brickwork in reservoirs should always be laid in Portland Cement or the best hydraulic lime.

Concrete is extensively used in waterworks construction, both for foundations and for entire structures. The materials for concrete are: (1) The aggregate or body. (2) The matrix or mortar.

The aggregate may consist of broken stone, slag, bits of brick, or almost any hard material. It should be broken to about a 2½-inch gauge. The matrix is either lime or cement and sand. The proportions in which the aggregate and matrix are taken should depend upon the proportion of void to solid in the former. This can be found out by filling a water-tight box of known capacity with the aggregate, and then noting the quantity of water that can be poured into the box without overflowing.

One cubic yard of stone, broken to 2½ inch gauge, contains 10 cubic feet voids; one cubic yard of ditto, broken to 2-inch gauge, contains 10⅔ feet voids; one cubic yard of ditto, broken to 1½ inch gauge, contains 11⅓ cubic feet voids.

Shingle contains 9 cubic feet voids. Thames ballast (which contains the necessary sand), contains 4 cubic feet voids.

If the aggregate consists of stones of various sizes, the voids will be reduced. When the concrete is intended for foundations where strength is necessary and imperviousness is immaterial, the matrix may be slightly less than the voids. If, however, imperviousness is the first consideration the matrix must *exceed* the voids.

The following is an estimate of the cost of concrete for the work above referred to:—

Materials for 1 cubic yard of concrete (6 parts broken stone to 1 part of mortar): broken stone, 6 parts = 27 cubic feet; sand, 2 parts = 9 cubic feet; Portland cement, 1 part = 3·51 bushels.

PARTS OF A DAY OCCUPIED BY A BRICKLAYER'S
LABOURER (HURST'S HANDBOOK)—

					Per cubic yard.
Measuring the materials	...	...	...	...	·04
Turning over twice	...	...	...	...	·06
Filling into barrows	...	...	...	...	·05
Wheeling, say 25 yards	...	...	...	...	·03
Levelling in layers	...	...	...	...	·02
Ramming	...	...	...	...	·03
					·23

ESTIMATE PER CUBIC YARD OF CONCRETE.

				s.	*d.*
Broken stone (2½-inch gauge), 27 cubic feet at 2s. 6d.				2	6
Sand, 9 cubic feet at 17s. 6d. per ton	...	...		7	10½
Portland cement, 3·51 bushels	...	...	...	13	2
Labour, ·23 days at 2s. 11d.	...	...	...	0	8½
				£1 4	3

The concrete should be mixed dry on a wooden platform, the materials being measured by means of wooden boxes without bottoms, turned over twice dry, sprinkled with sufficient water through a rose, turned over until thoroughly mixed, filled into wheelbarrows, wheeled to the site, tipped gently into position, and well rammed in 12-inch layers. One cubic yard is sufficient for one mixing.

Cement.—The cement most used in waterworks construction is Portland cement. It is used for making mortar, concrete, and for rendering.

The following estimate in connection with the reservoir above referred to for rendering may be useful. Cost per superficial yard of rendering on brickwork, ¾-inch thick, 1 of Portland cement to 2 of sand—

			s.	*d.*
Portland cement, ·21 bushel at 3s. 9d.	...	...		9½
Sand, ·42 at 1s. 1d.	...	...	...		5½
Plasterer and labourer, ·08 days at 14s. 7d.		...	1	2
			2	5

Portland Cement is grey in colour, weighs from 112 lbs.

per striked bushel, and should be ground so fine that after passing through a sieve containing 2500 meshes to the square inch, the residuum shall not exceed 10 per cent. Briquettes made from the cement and immersed in water, when sufficiently set, for seven days, should be capable of sustaining a tensile stress of at least 300 lbs. to the square inch.

Lime.—Hydraulic limes or limes capable of setting under water, are frequently used in waterworks construction, either alone or mixed with cement. The blue and brown Lias limes are examples. If of good quality, they give excellent results.

Gravel is used for filter-beds and for making concrete. In either case it must be clean and free from earth or vegetable matter.

Sand is used for filter-beds, and for mixing with cement and lime for mortar, and for making concrete. For filter-beds, the sand must be clean, uniform, but not too fine in grain, sharp, and approaching pure silica as closely as possible. For mortar and concrete the sand should be perfectly clean, free from clay or other impurities ; the grains should be sharp and angular.

Clay is largely used for puddling. It should contain only a small proportion of sand, and should be quite free from vegetable matter, or friable stone ; but the presence of a small amount of gravel gives it greater stability. The clay should be freed from all vegetable matter, and should be exposed to the weather or "weathered" for as long a period before use as possible. It should then be spread out flat, and cut across in every direction and thoroughly worked with spades, or passed through a pug-mill, sufficient water being added and the whole mass reduced to a stiff homogeneous consistency. It should then be rammed into position in very thin layers.

Wood is used in roof trusses, and in the construction of floors, doors, etc., in various parts of the work. It is also used for pile foundations. Oak, beech, and elm are the best suited for the latter purposes.

Innumerable tables have been published giving the relative strength of various materials to support loads. These strengths should never be approached in actual practice, and large margins should be allowed in designing structures from them. The following table of "Factors of Safety" is given in Unwin's "Machine Design."

		Dead load.	Live load.		
			In temporary structures.	In permanent structures.	In structures liable to shocks.
Wrought-iron	...	3	4	4 to 5	10
Cast-iron	...	3	4	5	10
Timber ...	...	—	4	10	—
Brickwork	...	—	—	6	—
Masonry	...	20	—	20 to 30	—

CHAPTER XIII.

STORAGE OF WATER.

ONE of the most important questions for the water engineer is the determination of the capacity for storage which is to be provided in the impounding reservoir. The purpose of these reservoirs is to maintain a balance between the fluctuations of supply and demand, when the rate of consumption is greater than the natural supply at the same period. They should not be too large or expensive, keeping in view the average growing necessities of the population. The average quantity of water required per head of population per day varies according to circumstances. It is generally supposed that it should not exceed twenty gallons in non-manufacturing, and thirty gallons in manufacturing towns; though, in fact, these quantities ought to be, and soon will be, regarded as maximum rather than minimum limits. The following table gives the variation in the quantity used in fifty-eight towns in the United Kingdom and in fourteen towns abroad. The comparison is not altogether satisfactory, as many of the towns abroad include water used for flushing purposes in the rate consumed per head :—

TABLE OF WATER CONSUMPTION PER HEAD PER DAY IN VARIOUS TOWNS IN 1892.

Town.	Galls. per head per day.	Town.	Galls. per head per day.
Aberdeen	60·0	Bath	21·5
Abingdon	5·0	Bedford	25·0
Banbury	17·0	Birmingham	23·0
Barrow-in-Furness ...	32·8	Blackburn	27·2

Town.			Galls. per head per day.	Town.			Galls. per head per day.
Bolton	...	...	20·0	Leamington	...	...	17·0
Bournemouth	...	...	25·0	Leeds	...	...	30·0
Bradford	...	...	22·7	Lincoln	...	...	21·0
Bridgwater	...	...	16·0	London County	...	...	29·5
Burnley	...	...	22·09	East London	...	...	31·8
Cardiff	...	...	21·0	New River	...	...	28·5
Carlisle	...	...	22·0	Chelsea ...	...	...	34·6
Carnarvon	...	...	40·0	West Middlesex		...	29·2
Congleton	...	...	10·0	Grand Junction		...	33·3
Coventry	...	...	20·0	Vauxhall Southwark		...	31·2
Croydon	...	...	32·0	Lambeth	...	...	30·4
Dartmouth	...	...	12·0	Manchester	...	...	21·0
Darwen	...	...	27·0	Newport (Mon.)	...	...	20·0
Derby	...	...	20·0	Northampton	...	...	14·0
Doncaster	...	...	20·0	Nottingham	...	...	18·73
Dover	...	...	26·0	Oldham	...	...	20·0
Dublin	...	...	47·0	Perth	...	...	39·39
Dundee	...	...	50·0	Ripon	...	...	23·0
Edinburgh	...	...	40·0	Salisbury	...	...	40·0
Glasgow	...	...	50·0	Sheffield	...	...	21·0
Halifax	...	...	23·75	Southampton	...	...	30·0
Hereford	...	...	30·0	Staleybridge	...	...	21·0
Huddersfield	...	...	22·5	Stratford-on-Avon		...	19·5
Keighley	...	...	30·0	Ulverston	...	...	40·0
Lanark	...	...	40·0	Warrington	...	...	20·0

TABLE OF WATER CONSUMPTION PER HEAD PER DAY IN FOREIGN TOWNS IN 1892.

Town.			Galls. per head per day.	Town.			Galls. per head per day.
Bayonne	...	...	55·0	Kiel	...	...	28·0
Berlin	...	...	22·9	Limoges	...	...	52·8
Bonn	...	...	63·0	Magdeburg	...	...	29·7
Boston, U.S.A.	...	...	76·0	Marseilles	...	...	99·0
Chicago	...	...	95·0	New York	...	...	65·0
Detroit	...	...	126·0	Paris	...	...	47·0
Frankfort	...	...	39·0	Philadelphia	...	...	56·0
Hamburg	...	...	52·0	Stuttgart	...	...	23·8

The consumption having been estimated according to the circumstances existing in the area to be supplied, the rainfall within the proposed catchment basin is then to be determined. In estimating the storage required, the data afforded by any single or average year will not be sufficient. The estimate must be based on a period of years during which the rainfall is below the average. There are several methods for determining the capacity of storage required, some by empirical formulæ, others by graphical methods. Extreme care should be taken in adopting empirical formulæ that due consideration is paid to the geological and other conditions existing within the catchment area to be dealt with. The late Mr. Thomas Hawkesley, F.R.S., deduced the following formula based upon his extensive experience—

$$z = \sqrt{\frac{1000}{r}}$$

Where z = the number of days' storage required (which varies from 100 to 250 days).

,, r = average rainfall in inches during three consecutive dry years, the average rainfall for a dry year being taken at five-sixths of that for an average rainfall of a long series of years.

A graphical method was communicated to the Liverpool Engineering Society, in December, 1891, by Mr. T. Turner Tudsbury, and a further paper was recently read before the Austrian Society of Civil Engineers, by Herr W. Rippl, of which the following is a description :—

On an axis of abscissæ, the months are laid off for each year of the period under consideration, and the demand of the town in cubic feet or gallons for each month plotted as ordinates, a slightly undulating curve drawn through the points so plotted gives the demand curve. In a similar way the supply curve is plotted, and represents the available rainfall from the water-shed.

Whenever the supply curve rises above the demand curve, we have a surplus on hand, and when the latter curve rises above the former a deficiency is shown for the period indicated. From such a diagram the surplus or deficiency for each month can be scaled off and used in the construction of a mass curve. The months and years of the period involved are represented by abscissæ as before ; but the ordinate at each month represents the algebraic sum of all the surpluses and deficiencies from the beginning of the period to that point. The mass curve reveals the surplus or deficiency during the interval between any two points on the axis of abscissæ, which is represented by the difference of the corresponding ordinates. An ascending curve shows an increasing, and a descending curve a decreasing storage; while crests and hollows show occasions when demand and supply are balanced.

IMPOUNDING OR STORAGE RESERVOIRS.

The principal factors which determine the position of impounding reservoirs are—

1. Purity of source.
2. Area of catchment basin.
3. Quantity of available rainfall.
4. Altitude and suitability of site.
5. Geological structure.

1. Purity of source, which has been already referred to, is of paramount importance. The most satisfactory area to impound water is where there is a sparse population, scant herbage, and an entire absence of cultivated land. The nearer these conditions are approached, either naturally or artificially, by collecting and diverting the sources of pollution, the nearer will an ideal water-shed be realized.

2. Area of catchment basin is found by drawing a contour line from the proposed site of the embankment along the ridges which form the water-shed, or from which the water sheds itself on either side. This may be ascertained from a contoured Ordnance map, or by running a contour on the ground and marking it on the map. The land within this line is then measured to arrive at the superficial area.

3. Quantity of available rainfall has been already described.

4. Altitude and suitability of site. The height of the source above the highest point of supply can easily be ascertained by an inspection of an Ordnance map, and a simple calculation will determine the available head for

supply. The pressure in the district to be supplied should at all times be sufficient to force the water above the highest buildings, and for this purpose an average pressure of 80 to 100 feet would be ample, and meet all requirements. In exceptional cases the only available catchment area is not at a sufficient elevation to supply the district, and then the water has to be pumped from the impounding reservoir to another reservoir at a higher elevation. The greatest capacity with the least cost is the next object to be attained. Where a valley or lake can be utilized, the most economical method of forming a reservoir is to construct a dam at the inlet, the most desirable shape (Fig. 16A) being where the valley gradually

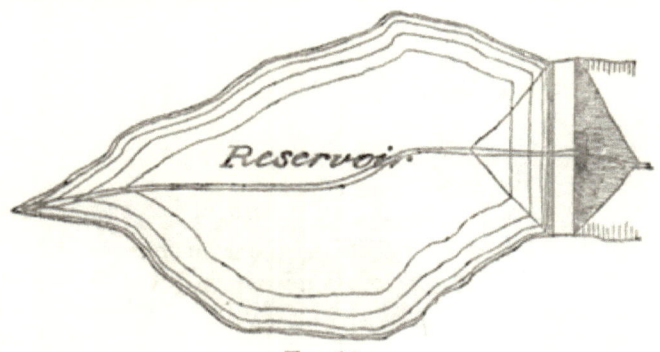

Fig. 16A.

widens out upwards from the dam with only a slight fall to it from the inlet. Such an arrangement would give the greatest capacity with the least cost for embankment, and the uniform depth resulting would prevent the growth of vegetation. Surface springs should be kept clear of the embankment on the inner side if possible, and if existing on the site of the outer portion, should be conveyed in pipes or concrete channels beyond the toe of the embankment. The materials of which the dam is to be formed will depend upon those which are available on the site of the dam, and will be dealt with in detail subsequently.

5. Geological structure of the area proposed to be utilized

for the formation of an impounding reservoir should be exceedingly carefully examined. Too much stress cannot be placed upon a thoroughly practical examination by an expert geologist or engineer who has a practical knowledge of the subject, as a want of sufficient care at the outset may result in a largely increased cost of construction, or even a subsequent abandonment of the site. The site of the Woodhead reservoir of the Manchester Corporation had to be abandoned on account of the unsatisfactory foundations revealed after the excavations had commenced. The presence of permeable rocks dipping towards the dam is a source of trouble, and every precaution should be taken to prevent the water from percolating into them. The best

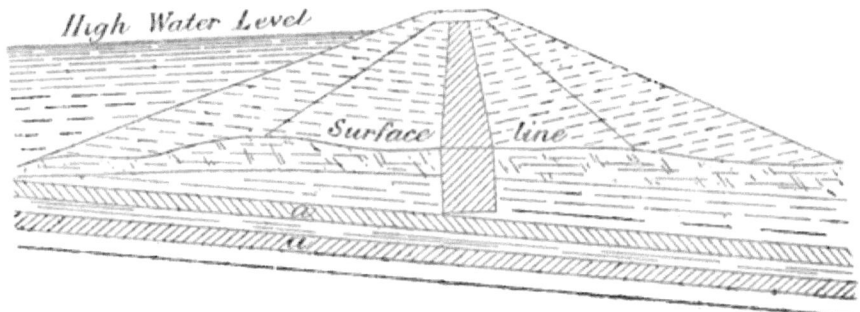

Fig. 17.   *a,* Pervious beds.

method is to excavate to a sufficient depth along the outcrop, and afterwards to fill in the trench with puddle protected on the surface with concrete. The permeable strata should also be cut through by the puddle trench of the dam, or the water in the reservoir will gradually escape and rise as springs at a lower point in the valley, the quantity varying with the head of water in the reservoir (Fig. 17). Mr. Isaac Roberts, F.G.S., records the following observations upon the effect of pressure on the quantity of water that will pass through a square foot of sandstone of average coarseness, $10\frac{1}{2}$ inches in thickness :—

Pressure.			Percolation.
10 lbs. per square inch	=		$4\frac{1}{2}$ gallons.
20	„	„	= $7\frac{1}{2}$ „
46	„	„	= 19 „

The dislocations produced by faults, and the fissures proceeding from them, are also a fruitful source of anxiety. A knowledge of their positions can only be ascertained by trial pits systematically arranged; boreholes should not be relied upon. The fault may be practically an open fissure, as well as a dislocation of the strata, raising a permeable bed to or near the level of an impervious one (Fig. 18). The

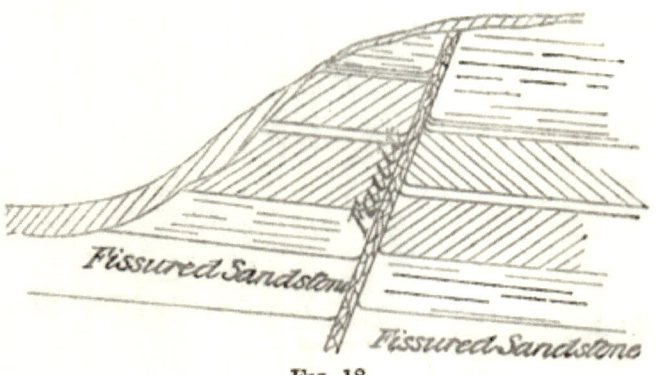

Fissured Sandstone

Fissured Sandstone

Fig. 18.

fissures extending on either side are even more difficult to deal with than the fault itself (Fig. 19); they frequently occur in the older rocks. The result of such fissures is to convey the water out of the reservoir either by the side of the dam or under it through permeable strata, where it may rise as springs. This entails a serious loss to the impounding works. In exceptional cases the matrix or material between the two cheeks of the fault is composed of fine silicious clay, which forms an effective dam in itself. The gorges forming outlets to valleys are frequently fissured, in many cases to such an extent as to render a site higher up the valley with a longer embankment preferable on the ground of economy of construction. The puddle trenches in many instances have to be carried to a depth of 150 to 200 feet, owing to the fissures found in the neck or gorge of a valley, the site of which superficially possesses an exaggerated importance owing to the short length of embankment required. The economy in one direction is, however, far exceeded by the costly foundations in the other. As an instance of serious results arising from a fissured foundation, where due precautions have not been taken, we may refer to the Holmfirth reservoir, which burst in 1852, on the only occasion on

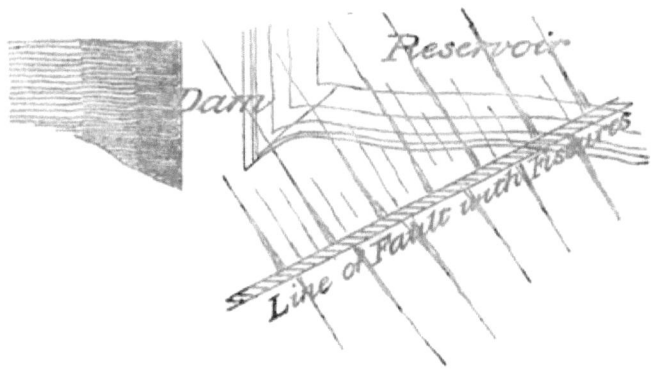

Fig. 19.

which it was filled. The embankment was constructed on fissured sandstone, and the water gradually escaped through the fissures, washing a portion of the embankment with it. Ultimately the embankment subsided below the weir-level, and a flood occurred completing its destruction.

CHAPTER XIV

THE site of the embankment, or dam, having been thoroughly proved by means of trial holes at least 5 feet square, and of sufficient depth to admit of a proper examination of the strata, the next step is to determine whether it is to be constructed of earth or masonry. Where there is a good compact clay foundation, and the clay is abundant in quantity, the dam must, for economical reasons, be formed of that material. Where, however, the position and quantity of suitable rock make conditions favourable for the construction of a masonry or concrete dam, then undoubtedly such a dam would be better, although the comparative cost would be much greater. Having decided upon the material for construction, it must be disposed of in the design according to experience, the recognized laws of such structures, and the peculiar circumstances of the case.

Earthen embankments, as employed in the storage of water, consist, as a rule, of two trapezoidal-shaped figures formed of earth, clay, and stone, supporting a centre core of puddled clay, increasing in width directly as the depth (Fig. 17). The proportions of earthwork dams are limited by the angle of repose or slope at which the materials employed will stand. With cohesive materials this depends upon their power of absorbing water, which can best be found by experiment. Experiments upon several clays used in reservoir embankments show that the absorption by weight varies from 12 to 53 per cent. In the latter case the embankment failed several times during construction.

In practice the outer slope should not be less than the ratio of 1½ horizontal to 1 vertical, and, as a rule, it is made either 2 or 2½ to 1. The inner slope, which has a greater tendency to slip, owing to its angle of stability being reduced by the water, should not be less than 2½ horizontal to 1 vertical, and is more frequently made 3 to 1.

The total width of the bank at the level of the top of the puddle-wall should not be less than three times the width of the puddle at that level. This width from slope to slope varies from 10 feet to 30 feet.

The height of the embankment above high-water level varies according to circumstances. In numerous cases where the inner slope is continued up to the top of the embank-

FIG. 20.

ment (Fig. 20) a greater height is required to prevent the waves from being driven over the top of the embankment in stormy weather. The late Mr. T. Stevenson, P.R.S.E., gives the following formula founded on his experience as a harbour engineer for finding the height of the waves in violent squalls :—

$$H = 1.5 \sqrt{D} + (2.5 - \sqrt[4]{D})$$

where H = the height of the waves in feet, where D = fetch in miles, which is the longest straight line that can be measured from any part of the dam to any part of the reservoir, when the latter is full and overflowing.

It is found to be more convenient to make the slope steeper above the water line with a storm-wall at its summit (Fig. 21), or to build a storm-wall entirely across the embankment at the high-water level, with a coping projecting at least 6 inches (Fig. 22). This has the effect of curving the waves back, and affords every protection to the top of the bank. The height of the top of the bank above the water in flat slopes should not be less than 8 feet vertical, and with

H

steeper slopes 5 feet vertical, with a dwarf wall at the water line not less than 4 feet vertical.

The width and batter or taper of the puddle-wall varies

FIG. 21.

with the materials of which it is composed. The clay used for the puddle should be carefully selected, and be of good

FIG. 22.

tenacious quality, comparatively free from sand, and entirely free from friable stones and vegetable matter. A small proportion of gravel is an advantage, and increases its stability. The clay should be turned over and weathered for two or three months, and then well cut, tempered, and worked in stages, and afterwards passed through pug-mills. It is then conveyed to the trench and inserted in layers. The top width of the puddle wall varies from 3 feet to 10 feet, and tapers outwards at from 1 in 8 to 1 in 16 down to the surface level, where it is either keyed into a concrete shoe as a base (Fig. 23) or is continued down in a trench until a sound foundation and retentive material are reached. The trench puddle is frequently carried down at a reverse or inward batter at rates of 1 in 8 to 1 in 16, according to circumstances (Fig. 24). When the pervious strata extend to a considerable depth it may be necessary to carry the trench down vertically the full surface width of the puddle-wall. Where this is done the bottom of the trench is covered with a layer of cement concrete connected to a key-piece of the same material, the layer at the base being

12 inches or more in thickness. Where the strata are very porous or fissured, a wall of concrete, stone, or brickwork,

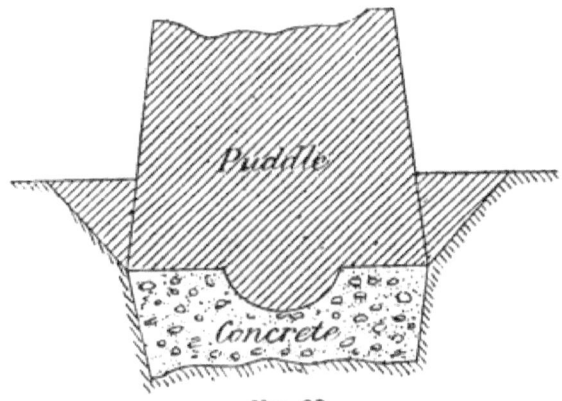

Fig. 23.

should be extended up the inner face of the trench as a protection to the puddle, and in some cases it is necessary

Fig. 24.

to use concrete in the trench instead of puddle. Water must in no case be in direct contact with the puddle.

The whole of the soil and earthy materials as well as

tree stumps and vegetation should be cleared off the site of
the embankment, and no vegetable earth used in the construction of the inner bank.   The materials in the inner portion
should consist of fine clayey or other adhesive material with
a small proportion of stones or ballast, except towards the
toe, where the proportion of stone should be increased, the
outer portion consisting of dry, hard, and stony materials,
with dry stone drains where necessary.   The materials on
either side of the bank should be well consolidated as the
work proceeds.   On either side of the puddle-wall a
width of selected clayey material of not less than four times
that of the puddle wall is formed for the purpose of
keeping the puddle moist, and to assist in its protection.
The whole of the materials should be deposited in layers of
from 9 inches to 2 feet in thickness, curving or dipping
towards the puddle-wall on either side.   In some cases a
bed of puddle is carried from the wall under the base of the
inner slope and continued up the slope, sufficiently protected
with selected material.   The object of this is to render the
inner bank impermeable (Fig. 25).

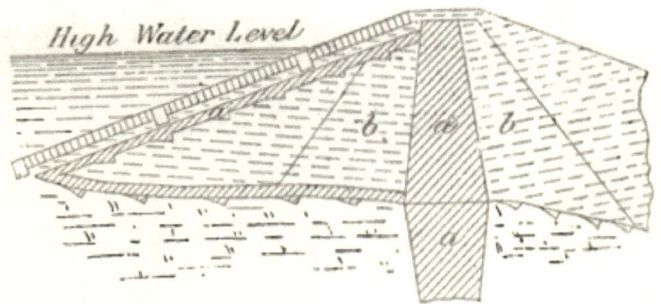

FIG. 25.

a, Puddle ; *b*, selected material.

In excavating within the reservoir area for the purpose of
providing material for the construction of the embankment,
or with a view to increasing the capacity of the reservoir,
care must be taken not to remove an impervious covering

over pervious strata, and thus create a difficulty which it should be the main object to avoid.

The inner slope should be protected with stone pitching over the entire area of the made embankment after it has become consolidated, the toe of this pitching being embedded in a concrete footing. The solid slopes, when of a clayey nature, are also usually pitched with stone for a vertical height of 3 feet above and at least 5 feet below the high-water line, as a protection from the wash of the waves, and as a preventative against the growth of vegetation in the shallow water, as well as against discoloration from the dissolved clayey matter.

In cases where a reservoir is constructed practically on a table-land and embanked all round, sufficient material is excavated from the interior to form the embankments and provide the requisite capacity. The methods of construction are in every way similar to the foregoing.

There are two indispensable accessories to an impounding reservoir, viz. the outlet and overflow weir, over which many difficulties have arisen and through which many disasters have occurred.

The outlet arrangements are carried out in several ways, according to the special circumstances of each case. The method of carrying the outlet pipes through the deep portion of the made embankment has been rarely followed, and is only permissible in shallow reservoirs. The terrible disaster in 1864, at the Bradfield or Dale Dyke Reservoir, near Sheffield, when 250 lives were lost, resulted from this practice. In reservoirs not exceeding 25 feet in depth a syphon (the action of which has already been explained), is the most economical and efficient as well as the safest method of drawing the water off. It does not interfere in any way with the embankment below the high-water line, and the same method has been recommended by Sir Robert Rawlinson for drawing off the lower water from large reservoirs (Fig. 26). This obviates the necessity of carrying the tunnel outlet at so low a level. The advantages of doing so are less interference with the strata at great depths and economy in construction. The system generally adopted in large reservoirs is to cut a trench or drive a heading through the solid rock at one end of the embankment, and construct a

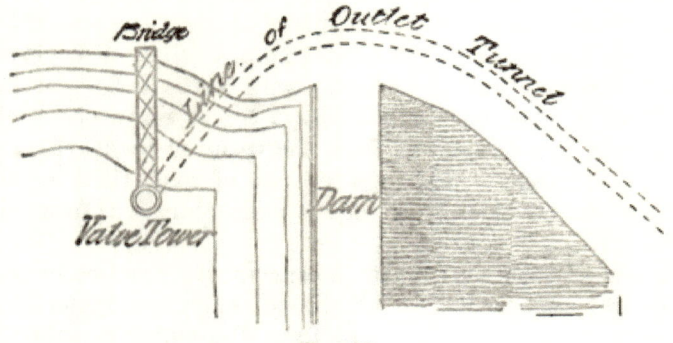

Fig. 26.

stone, brick, or concrete and iron culvert with a valve tower in the reservoir (Fig. 27). The base of the tower is below the deepest portion of the reservoir, unless a syphon 'is

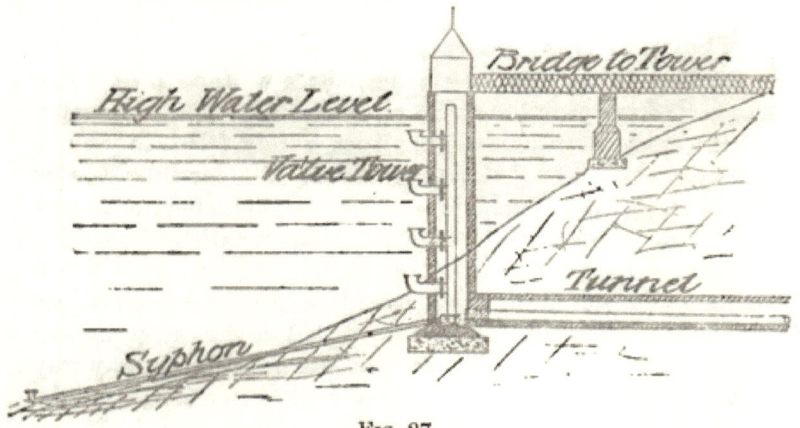

Fig. 27.

adopted and arranged to draw off the water at different levels. The water is discharged through the supply main laid within the culvert. The advantage of this system is the facility with which any of the working parts can be examined and repaired without interfering with the embankment or being in any way a source of weakness to it. The question whether it is preferable to drive a heading or cut an open trench for the culvert depends upon the nature of the rock, apart from economical reasons. Where the rock is solid and compact a heading is preferable in most cases, but where the rock is fissured and contains many "backs," an open trench, which is filled in as the culvert progresses, is frequently the better course. This is due to the considerable difficulty attendant upon consolidating around the culvert in a timbered heading in fissured ground. In cases where it is necessary to cross the puddle-trench at a higher level than the bottom of the trench, the culvert should be

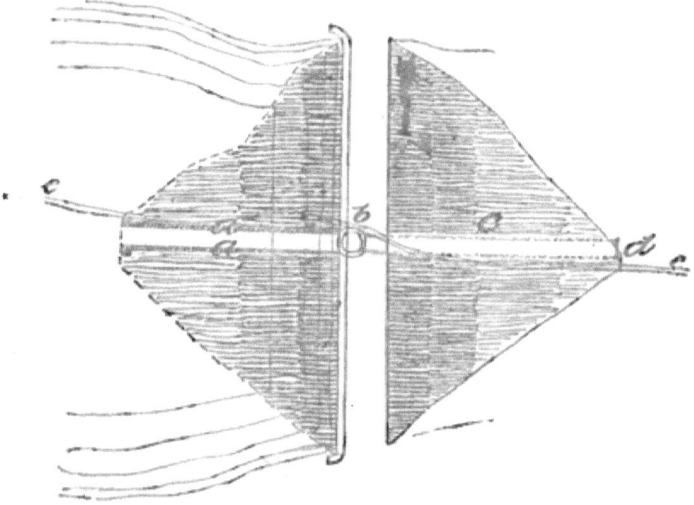

Fig. 28.

a, Wing walls; b, valve well; c, culvert; d, supply main;
e, course of stream.

supported on a concrete pier brought up from the solid rock. The valve tower may either be constructed of stone, brick, or iron, with draw-off pipes at different levels, communicating with a stand-pipe in the centre of the tower. Each draw-off pipe is controlled by a valve worked from the top of the tower, and the bell-mouths are turned upwards so as to admit of plugging from the surface in the event of anything going wrong with the valves within the tower.

The method represented by Figs. 28 and 29, is not a system to be recommended. Here the outlet is placed in the deepest part of the embankment, and consists of a

FIG. 29.

a, Masonry forebay; *b*, valve well; *c*, culvert; *d*, supply main.

masonry forebay, supported by iron struts, and a draw-off well and tunnel, also in masonry. The tunnel is supported on a concrete pier (with or without slip-joints), where it crosses the puddle-trench. This system is objectionable, from the fact that many of these tunnels have been distorted or cracked, frequently developing leaks. Such structures, in all probability, constitute an element of weakness where the greatest strength is required.

CHAPTER XV

STORAGE OF WATER—*continued.*

THE overflow or waste weir is placed at the embankment ond of the reservoir on solid ground, with a concrete foundation. It consists of a heavy masonry base formed of large stones set in cement and well keyed together, with heavy pitching on the approach and discharge sides. The channel

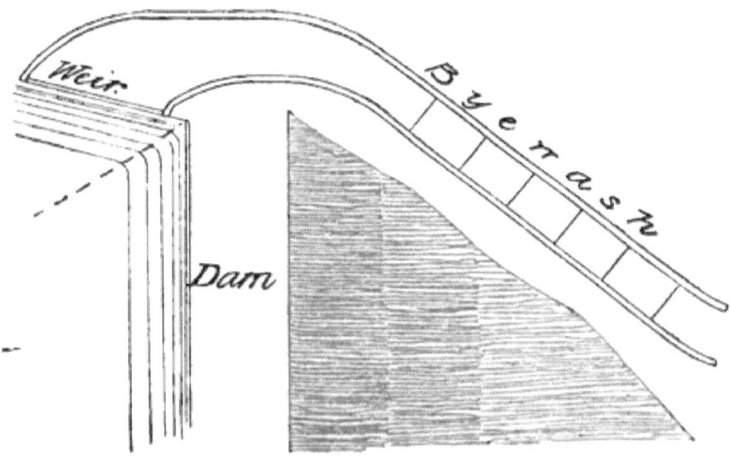

FIG. 30.

on the discharge side is continued in a series of steps down the side of the embankment (Figs. 30 and 31) (the watercourse being pitched with stone), and terminates in the original stream-course of the valley clear of the embankment. The purpose of the overflow weir is to prevent the

water in the reservoir from rising above the level of the
embankment and flowing over, and, in the case of earthen
structures, causing the inevitable destruction of the works.
In practice the length of the weir is made from 2½ feet to
4 feet per 100 acres of water-shed. The length is limited
by the maximum height to which the water is allowed to
rise above the crest of the weir, which should never exceed
2 feet, and is generally fixed at 18 inches. The conditions
peculiar to each gathering-ground must be taken into consideration in the design of such works, but the rules given
above may be safely followed where no storm records are
available. An instance of insufficient length of weir in connection with the Tittesworth reservoir of the Potteries

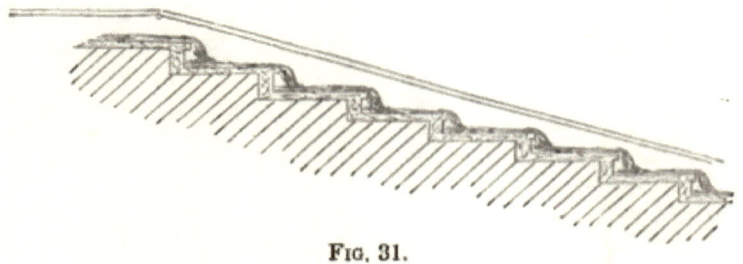

Fig. 31.

Waterworks occurred in 1862. In this case the drainage
area was 6800 acres, and the waste-weir was 60 feet in
length. The water in the reservoir rose 5 feet above the
crest of the weir, and to within 1 foot of the top of the
embankment. Where a bye-wash channel has been constructed round the margin of the reservoir, from the inlet,
of sufficient capacity, the length of the overflow weir may
be reduced accordingly. A residuum pond is frequently
constructed at the inlet end of a storage reservoir, with
considerable advantage. This has the effect of reducing
the velocity of the storm waters, arresting any detritus, and
allowing the water to deposit the greater part of the matter
held in suspension. The last is a matter of some importance where the storm-waters are exceedingly turbid. The

pond is formed by constructing a wall or embankment across the mouth of the inlet, the top level of which being about 12 inches above the high-water level of the reservoir. In the case of an embankment it is necessary to face the top and slopes with heavy stone pitching or concrete. The inlet water is allowed to rise over the top of the residuum wall (which forms one long weir), and fall into the storage reservoir. The pond may be cleansed either by drawing off the water and removing the deposit by manual labour, or by means of pipes connected with the pond and continued through the reservoir to its outlet, delivering a continuous stream of sludge-water into the original river channel below the embankment.

The design and construction of masonry or concrete dams, being rarely necessary for rural supplies, do not come within the scope of this book.

The cost of storage reservoirs with earthen embankments varies considerably, according to circumstances, from £70 to £900 per million gallons of capacity.

Service reservoirs are supplied direct from the impounding reservoir, or, where filtration is necessary, from the filter-beds. Their office is to regulate the variation in the daily consumption, and to provide sufficient storage to meet the requirements of supply in the event of any accident of a temporary nature occurring between them and the source. The quantity of storage to be provided varies according to circumstances; but, as a rule, two days' storage will meet all emergencies. Where the source of supply is at a considerable distance, or somewhat inaccessible, and where there is a single main, or, in the case of a pumping supply, where there are no duplicate arrangements, it would be prudent to increase the storage capacity of the reservoir so as to make provision in the event of a breakdown.

In order to enable the student to form an idea as to the variation during the day, the accompanying diagram (Fig. 32) is given, which is taken from a Deacon differentiating meter. This diagram shows graphically the daily variation in a manufacturing town of 48,258 inhabitants,

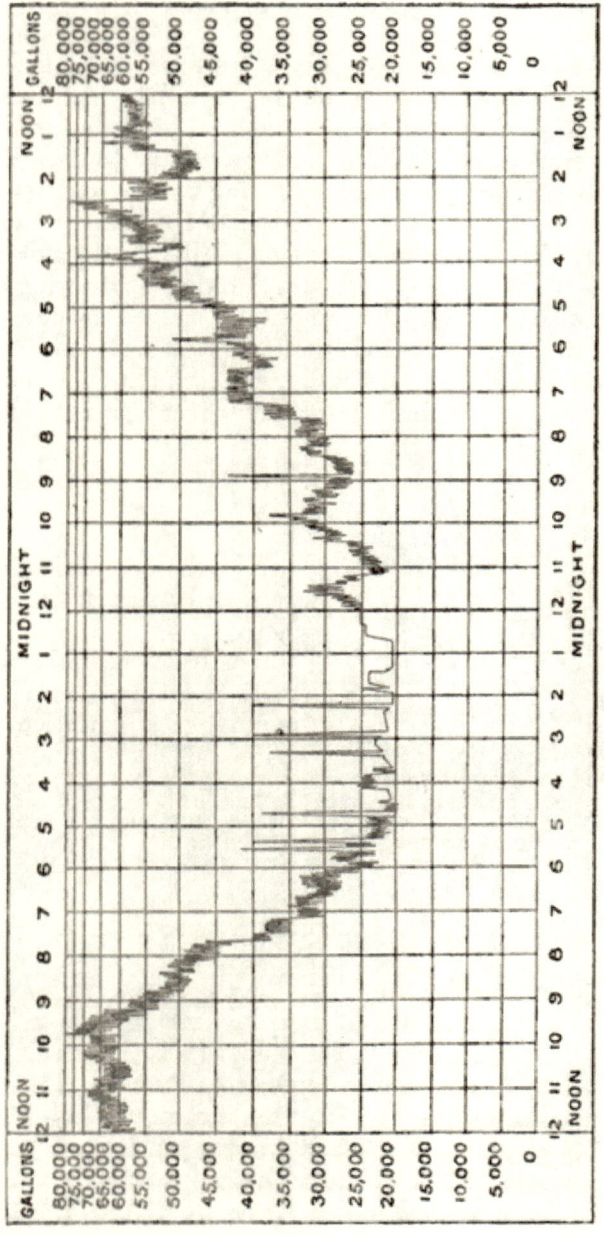

Fig. 32.

and may be taken as a fair example. The abrupt rises and
falls shown on it between 2 a.m. and 6 a.m. are due to the
supply to locomotives during those hours, the waste line
or minimum flow being equal to 20,300 gallons per hour,
which is chiefly due to defective fittings. The total consumption, including domestic and trade supply, per head
of population is equal to 19·85 gallons, and the minimum
flow 10·09 gallons per head, or practically one-half of
the supply.

The variations of supply during the different periods of
the year are not so great as might be anticipated. The
householder's favourite practice of allowing the taps to
run during frosty weather, and the number of burst pipes,
have the effect of raising the consumption in the town
referred to frequently up to, and in excess of, the summer
months. The following table gives the consumption per
head per day for the last four years :—

Gallons per Head per Day.

Month.				1890.	1891.	1892.	1893.
January	...	...	...	19·77	29·30	23·26	28·50
February	...	...	...	19·57	23·26	21·24	23·43
March	...	...	...	20·38	21·54	22·53	22·75
April	...	...	...	18·95	20·38	23·11	23·14
May ...	...	...	...	20·99	21·06	23·39	22·11
June...	...	...	...	20·81	22·50	23·31	23·25
July ...	...	...	...	21·05	23·62	22·90	21·69
August	...	...	...	21·79	21·66	21·73	20·63
September	...	...	...	21·40	22·13	21·72	18·78
October	...	...	...	21·10	22·13	22·08	20·08
November	...	...	...	20·58	21·30	21·26	21·91
December	...	...	...	24·21	23·54	23·29	19·90

The site for a service reservoir should be at a sufficient
elevation, and within the immediate vicinity of the district
to be supplied. It is usually constructed either of masonry,
brickwork, or concrete, and roofed over; or by excavation
and embankments lined with concrete and pitching and left

open. It is absolutely necessary to cover the service reservoir when near a town or manufactories to prevent contamination, and especially so after filtration. Service reservoirs are only left uncovered when situated at some distance from any smoke or fumes from chemical or other works, and in such cases the depth must not be less than 10 feet, which may be increased with advantage so as to

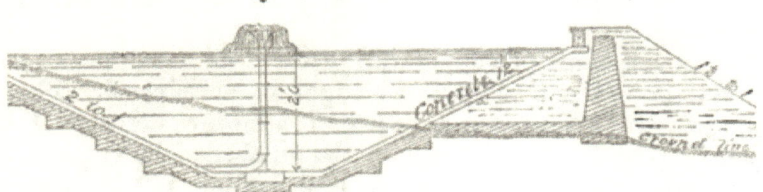

Fig. 33a.

prevent the growth of vegetable matter which produces that peculiar fish-like smell so common in shallow reservoirs. Figs. 33a and 33b are examples of open reservoirs, and Fig. 34 of a covered reservoir. Covered reservoirs should always have two feet of earth above the roof, to keep the water as cool as possible, and ventilators should be placed in the crowns of the arches. It is an advantage to have

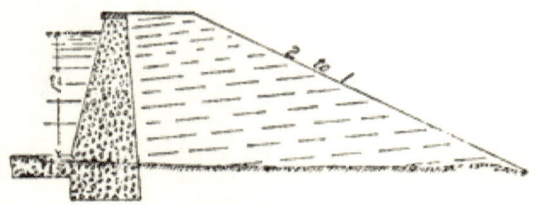

Fig. 33b.

a wall dividing the tank or reservoir into two portions for the purpose of cleansing from time to time.

Collecting tanks are used for storing the water from springs, and fulfil the offices of impounding reservoirs on a small scale, to which the duty of a service-tank is frequently added. These tanks are constructed of masonry, brickwork, or concrete, either with arched roofs, as in Fig.

34, or covered with iron plates supported by girders (Fig. 35).

The cost of covered service reservoirs varies from £2 to £6, and of open reservoirs from £1 per 1000 gallons.

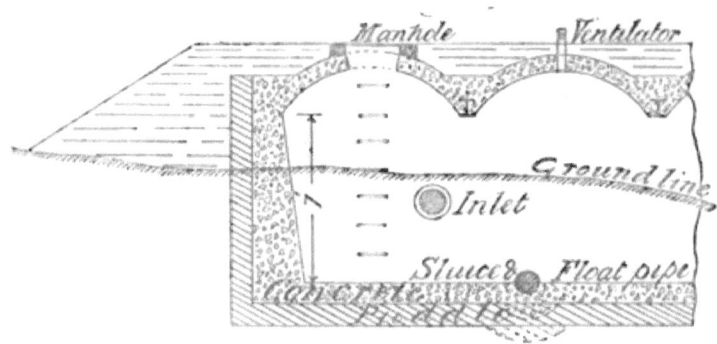

Fig. 34.

The usual accessories to a service or collecting reservoir, containing from 5000 galls. to 50,000 gallons, are the inlet and outlet, overflow and wash-out or scour-pipes. The inlet-pipe should be so arranged that the inflowing water may be shut off or diverted from the reservoir when the latter is being repaired, etc. The mouth of the inlet-pipe is usually fixed slightly above the level of overflow. The outlet should be a few inches above the level of the floor of the reservoir, so as to allow for a certain amount of deposit from the water. Its mouth should be covered with a perforated cap, rose, or strainer, which is best constructed of tinned copper. The outlet should be commanded by a sluice-valve, fixed inside the reservoir, worked from above by a wheel and spindle. The supply from a reservoir is sometimes taken by means of a floating pipe (Fig. 35). This ingenious method allows of the water being always taken from a little below the surface, which is the clearest portion of the water in a reservoir. The overflow pipe is either a pipe taken through the wall of the reservoir, with its mouth at the highest point to which the water is to be allowed to rise, or it may consist of a vertical pipe carried up from the floor of the reservoir, having a bell-mouth for receiving the overflow water, and an inlet at the base controlled by a valve which acts as a scour or wash-out pipe. In the latter case the pipe should be constructed of copper. The wash-out, or scour-pipe, has its mouth situated at the lowest point of the floor of the reservoir, which should be made to slope towards it. It should be large enough to empty the reservoir rapidly, and must have its outlet below

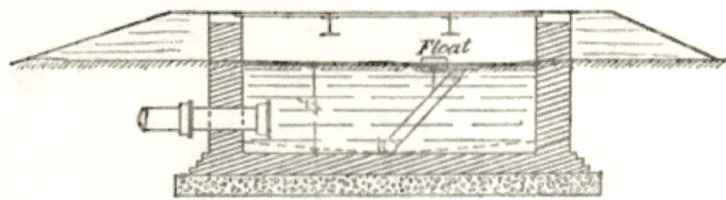

Fig. 35.

the level of the floor of the reservoir. It may either be controlled by a sluice-valve worked from the surface, or in small tanks by means of a brass plug and chain. In small tanks the overflow and wash-out are sometimes combined, the foot of the vertical overflow pipe being ground into the mouth of the wash-out; by loosening and lifting the overflow pipe the water is free to escape through the wash-out. Ladder-irons should be built into the wall of the reservoir to allow of access to the interior.

The inner surface of the reservoir should be rendered with cement, which should be brought to a perfectly smooth surface with the trowel.

Where the supply is obtained from springs, it is usually necessary to collect them by means of stone-ware pipes with open joints covered with broken stone. These pipes are connected by means of close-jointed pipes, and conveyed to a small tank, from whence they are conducted to the reservoir. Great care must be taken in collecting springs, to avoid all chance of pollution.

CHAPTER XVI

THE PURIFICATION OF WATER—FILTRATION.

THE methods of " purifying " or rendering water suitable for domestic consumption are aëration, subsidence, precipitation, straining, and filtration.

Aëration is a natural process of oxidation, the atmosphere acting on matter in solution, this action being facilitated by forming cascades and fountains to agitate and break up the water into thin sheets and spray. This method is employed by the West Gloucester Water Company, at Frampton Cotterill, to get rid of the large amount of dissolved sulphuretted hydrogen contained in the water, which by this means is rendered bright and more palatable. Exposure to the atmosphere has the effect of softening hard waters by releasing the loosely combined carbonic acid and precipitating the carbonate of lime, but in such cases there is great liability to develop vegetable growth. The beneficial effects of aëration through the use of fountain inlets in destroying algæ have been proved—in two instances with remarkable results. The action of the atmosphere on running streams in rivers and channels is well known, the organic impurities being brought in contact with the oxygen of the atmosphere, and gradually oxydized and rendered innocuous.

Subsidence is a process of settlement or gravitation of matter held in suspension, its rapidity depending on the specific gravity or fineness of the matter to be deposited. This action is continually proceeding in storage reservoirs to a greater or less extent, according to the condition of the water, as well as in the settling ponds, residuum lodges, and shallow reservoirs, which are specially adapted for the purpose, and are usually a preparatory stage for filtration.

Precipitation of certain impurities is produced by the addition of a precipitant, the most economical being caustic lime. A certain quantity of lime is added to a measured quantity of water in a tank, forming what is known as lime water; the clear liquid is drawn off by a float-pipe into another tank, and the water to be softened is added to it, the action being as follows : The caustic lime combines with the loosely combined carbonic acid in the water, forming carbonate of lime, which is precipitated along with the carbonates already in solution. The lime process was patented by Dr. Thomas Clarke, of Aberdeen, in 1841, and all the more recent methods are based on this principle. It has been successfully applied in several waterworks, both for domestic and manufacturing purposes, and among the towns using one or more of the recent methods may be mentioned Colne Valley, Southampton, Wellingborough, Saffron Walden, St. Helen's, and Stroud. The cost of softening, to the extent of removing from 10 to 24 degrees of hardness, varies from ¼d. to ½d. per 1000 gallons. The hardness of water is stated in degrees, each degree representing one grain of carbonate of lime per gallon, and is found by noting the quantity of standard soap solution required to produce a permanent lather in a gallon of water. The composition of clear lime-water being constant, it is found that if the degrees of hardness are divided into 130 it will approximately give the number of gallons of the water which can be softened by one gallon of lime water. The above process has the great advantage of destroying organic matter and producing a bright effluent. The following table gives the hardness of water in a large number of towns :—

Name of town.	Degrees of hardness.	Name of town.			Degrees of hardness.
Glasgow (Loch Katrine)	0·80	Dundee ...	...	...	3·28
Manchester (Thirlmere)	1·50	Bournemouth	...	...	4·70
Sheffield	1·50	Worcester ...	...	...	8·06
Liverpool (Lake Vyrnwy)	3·15	Lowestoft ...	...	...	9·00

Name of town.			Degrees of hardness.	Name of town.				Degrees of hardness.
Newport (Mon.)	...	...	10·40	London	...	...	...	16·00
Cheltenham	...	...	11·56	Portsmouth	...	...	...	16·00
Yarmouth	...	...	12·60	Canterbury	...	...	...	17·00
Bristol	...	...	13·40	Stroud	...	...	...	17·00
Northwich	...	...	13·60	Windsor	...	...	...	17·89
Nottingham	...	...	13·60	Southport	...	...	...	17·90
Newcastle-on-Tyne			14·00	York	...	...	...	18·00
Reading	...	...	14·50	Southampton	...	...	...	18·00
St. Helen's	...	...	15·00	Sunderland	...	...	...	24·00
Northampton	...	...	15·47	Wellingborough	...	...	...	37·00

The commercial and domestic economic advantages which a soft water possesses over a hard one are indisputable. The late Mr. Thomas Hawkesley, in recent evidence, however, stated that the death-rates for ten years, from 1882 to 1891, in twenty-seven large towns supplied with hard and soft water were: Hard-water supply, 20·2 per 1000 persons; and soft-water supply, 23·0 per 1000 persons = 13·9 per cent. excess over hard-water supplies.

Straining, *e.g.* through screens of brass or copper set in wooden frames—is absolutely necessary in all reservoirs. The screens intercept all floating and suspended matter larger than the mesh. They are removed from time to time for cleansing, which is usually performed by the application of a jet of water from a hose-pipe. The principle of filtration through sand, for the purpose of removing matters held in suspension, is often imperfectly apprehended, the popular idea being that the sand simply acts as a sieve, and prevents the passage of any particles larger than the interstices between the grains, at the same time allowing a certain amount of subsidence to take place upon the upper surfaces. The sand, however, does much more than this—the main action being due to the force of adhesion or mutual attraction between the particles in suspension and the whole surfaces of the grains of sand, and not the top surfaces only, as would be the case if the action were merely that of subsidence. It has also an effect, although small, on matters in solution, which is illustrated by the following analysis, by Dr. Percy Frankland, of river water before and after filtration :—

RESULTS OF ANALYSES EXPRESSED IN PARTS PER 100,000.

	Before filtration.	After sand filtration.
Total solid matters	28·40	26·20
Organic carbon	·123	·119
„ nitrogen	·025	·022
Ammonia	·0	·0
Nitrogen as nitrates and nitrites ..	·077	·089
Total combined nitrogen	·102	·111
Chlorine	1·6	1·6
Hardness, temporary	11·5	10·9
„ permanent	7·1	7·1
„ total	18·6	18·0

The fact that chemical analysis showed only a slight improvement in the water after sand-filtration somewhat threw discredit upon sand-filters, and it is only within the last few years, since the methods of Koch and others drew the attention of scientists to the bacteriological examination of water, that the remarkable efficiency attained by properly managed sand-filters in reducing the number of bacteria in water has been recognized. It was found that from 95 to 99 per cent. of the micro-organisms were removed by filtration from the London Water Company's supplies, reducing to a minimum the risk of pathogenic or disease-forming bacteria passing through the filters to the consumer. Dr. Percy Frankland has found that the water supplied to London after filtration contains less bacteria than many lake waters, a comparison of which is given as follows:—

New River (London), 38 colonies from 1 c.c. of water.

Grand Junction (London), 47 colonies from 1 c.c. of water.

Loch Katrine (Glasgow), 74 colonies from 1 c.c. of water.

Loch Lintrathen (Dundee), 161 colonies from 1 c.c. of water.

Lake Lucerne (Switzerland), 50 colonies from 1 c.c. of water.

Lake Geneva (Switzerland), 38 colonies from 1 c.c. of water.

Lake Constance (Switzerland), 58 colonies from 1 c.c. of water.

It would therefore appear that too much importance must not be attached to the number of bacteria present in drinking water, within certain limits, provided they are not of a pathogenic nature. The varieties of bacteria are very numerous, but most of these, with the exception of probably a few species, are beneficial rather than otherwise. Amongst the pathogenic bacteria that have been detected in water are the bacilli of tetanus, anthrax, typhoid, and the cholera spirillum. The advantages of sand filtration were strikingly illustrated at Hamburg and Altona during the cholera epidemic in 1892. These cities derive their supply from the river Elbe, the former without filtration, and the latter at a point in the river below the outfall sewers of both cities, but properly filtered, with the result that the relative proportion of cholera cases per 10,000 inhabitants was: Hamburg 290, and Altona 40 (of which many were imported cases).

A fact, which is receiving much attention from biologists at the present time, is that a filter-bed does not reach its normal state of efficiency, or technically "become ripe," until it has been in use five or six days; this is believed to be due to the formation on the surface of the sand of a gelatinous microbic tissue (zoogloea) produced by bacteria.

In the design of filter-beds many engineers take advantage of the site when on sloping ground to place the beds at different levels; others prefer to keep one level throughout by excavations and embankments. The area of each bed should be arranged so as to give an equal flow in the drains, but should not be excessive; and the distance for wheeling when the sand is being removed should not be too great. The number of beds should be sufficient to permit of half of them being out of use for cleansing purposes, the supply being maintained through the others. It is found convenient and economical to arrange the sand-washing apparatus in the centre of a battery of filter-beds. In some cases the washing apparatus is fixed in the centre of the bed, as at Belfast, but this arrangement is not generally adopted; another method, which is, without doubt, the proper one, is to periodically reverse the filters and allow the water to flow upwards, and thus carry off the impurities through an over-

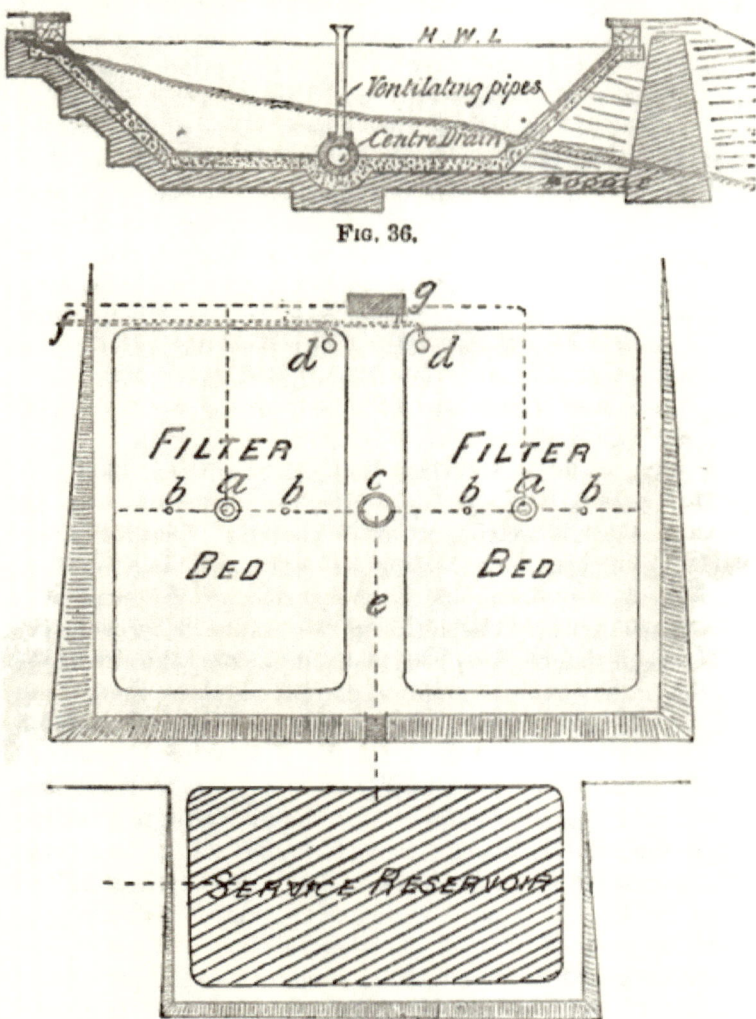

FIG. 36.

FIG. 37.

a, Inlet; *b*, ventilating pipes; *c*, valve well; *d*, overflow; *e*, filtered water main; *f*, overflow main; *g*, sand washer.

flow. Where this arrangement has been applied, the beds have attained their efficiency within twenty-four hours. The basin of the filter-bed is either constructed partly by means of excavation and embankment, with a puddle-wall

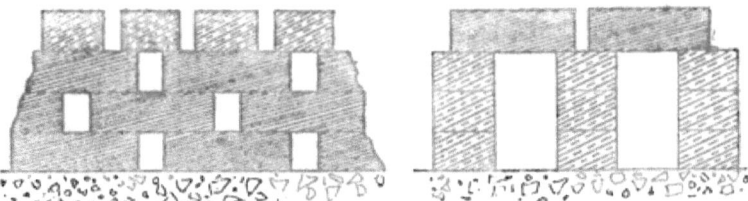

Fig. 38.

lined with concrete (Fig. 36), or with concrete walls backed up with earth (Fig. 33B, Chap. XV.). The floor of the basin is formed so as to dip towards the outlet of the filter, which communicates with a valve well, from which it is conveyed to a clear-water basin (Fig. 37). The centre

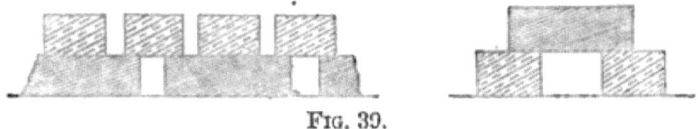

Fig. 39.

or main drain in the filter-basin is constructed of brick-work, concrete or perforated glazed pipes, and the side or arterial drains of perforated pipes or bricks laid dry with spaced joints (Figs. 38, 39, 40, 41). Ventilating pipes are carried up the slopes or side walls above the water-level

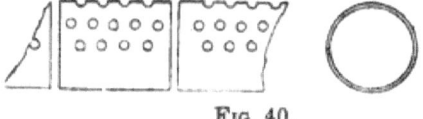

Fig. 40.

from each of the arterial drains, and from 2 to 4 on the line of main drain. The inlet is arranged in various ways. Fig. 42 shows an arrangement that has been adopted with great success.

The bed is formed of a layer of stone broken to pass through a 3½-inch. ring, but not through a 2½-inch. ring, and from 2 feet to 3 feet in thickness; this is succeeded by a layer of gravel 12 to 18 inches thick in two or three

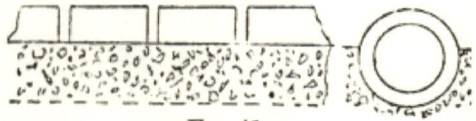

FIG. 41.

degrees of fineness, in some cases perforated tiles are used in preference; finally, the filtering medium of sand is spread over the whole of the supporting material (Fig. 45), the thickness varying at different works as shown in the following table:—

THICKNESS OF SAND-FILTERS.

	Maximum.		Minimum.	
	Ft.	Ins.	Ft.	Ins.
Chelsea, London	4	6	3	6
West Middlesex, London	3	3	2	6
Southwark, London ...	3	0	1	6
Grand Junction, London	2	0	1	3
Lambeth, London ...	3	0	2	6
New River, London ...	2	3	1	5
East London, London ...	2	0	1	4
Dublin	2	6	1	0
Bristol	2	0	1	0
Malvern	2	6	1	6
Harrogate...	2	0	1	0
Paisley	2	0	1	0
Barrow-in-Furness ...	2	0	1	0
Ulverston	2	0	1	0

Dr. Sims Woodhead, who has devoted considerable attention to the subject, suggests a minimum thickness of 3 feet. The sand for filtration should be hard and angular, and thoroughly washed, as well as the supporting material, before being deposited in the filter-basin. The rate of filtration should not exceed 5 inches per hour, or 2½ gallons

per hour per square foot. The mean rate of filtration in
the London filters is less than 2½ gallons per hour. The

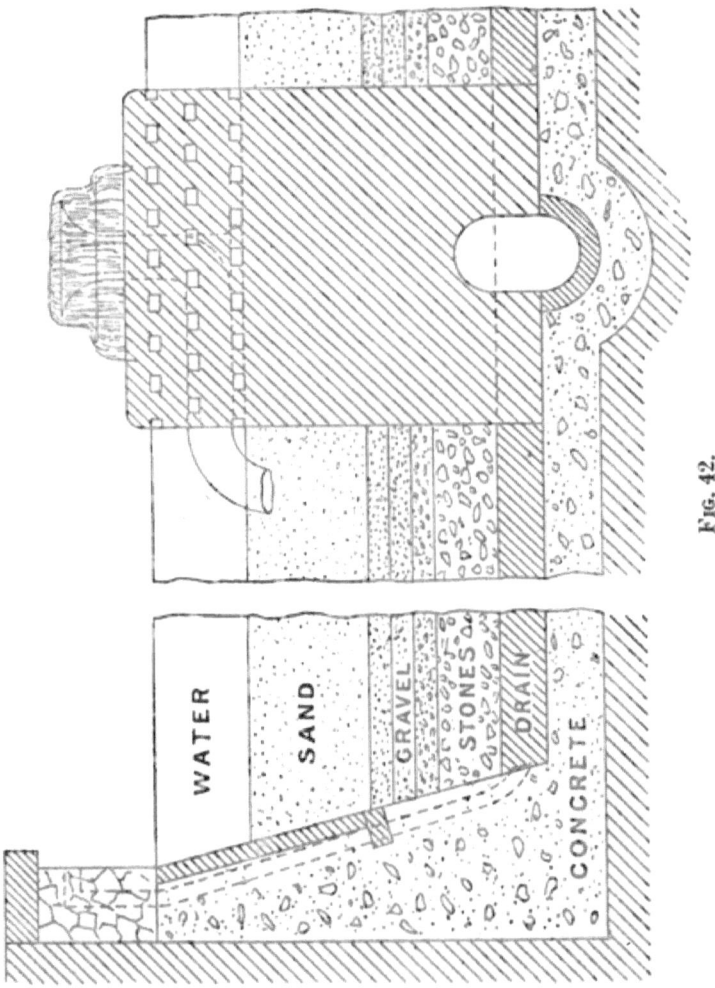

FIG. 42.

depth of water in a filter-bed in this country is usually
from 2 feet to 3 feet, and from 4 feet to 7 feet where
exposed to very low temperature, unless the beds are

covered over. The process of cleansing the filter-bed, where downward pressure only can be resorted to, is to remove a thin layer of sand half an inch or more in thickness, containing the perceptible suspended matter, and, if the sand is costly and has to be washed again for future use, to deposit it near the sand-washing apparatus, otherwise it is removed into a waste heap. The surface thus bared by the above process is raked over with a long pronged rake on two or three occasions, allowing a few days to elapse between each operation to aërate the sand.

The cost of constructing sand-filters per square yard varies from £1 to £4, according to circumstances, and the cost of filtration, exclusive of capital, is from 4s. 6d. to 7s. per million gallons. "Magnetic carbide," spongy iron, "polarite," and other media have been applied with beneficial results to the treatment of impure waters, producing a bright and pure effluent; and, in fact, with river waters the results fully justify the increased expenditure through their use. Mechanical filters, such as Dr. Anderson's Revolving Cylinders, containing oxide of iron, in a few cases where they have been employed, as at Hamburg, have given satisfactory results.

CHAPTER XVII.

PIPES.

WATER is conveyed in the various stages from the source to the consumer by means of open channels, tunnels, and culverts, or by pipes of cast- or wrought-iron, steel, lead, clay, and wood. Where the water is conveyed under pressure, pipes must be used, and they are generally more convenient and economical in construction. The thickness of the shell of the pipe and the form of joint depend upon the material employed, the capacity, and the pressure it will be required to withstand. In calculating the thickness of the shell, sufficient allowance must be made for imperfect workmanship, shocks in handling and laying in the trenches, also the weight of the superincumbent earth, and the traffic they will have to support, as well as the great strain which may come upon them on account of the sudden opening or closing of valves.

The bursting strength of pipes is found by the following formulæ :—

$$1. \quad p = s \times \text{hyp log R}$$

$$2. \quad s = \frac{p}{\text{hyp log R}}$$

$$3. \quad \text{Hyp log R} = \frac{p}{s}$$

Where p = the internal pressure in tons per square inch.

,, s = the maximum tensile stress in tons per square inch—7 tons being usually adopted as the value for cast-iron.

,, R = the ratio of the outside diameter to the inside diameter.

A high factor of safety must be used to adequately allow for the strains imposed on the metal. Empirical formulæ based on practice are found to be more convenient for the purpose of determining the thickness of metal. These formulæ are numerous and widely divergent in their results, but the following, suggested by the late Mr. J. La Trobe Bateman, has been found to work well in practice :—

$$t = \cdot 25 + \frac{Hd}{9600}$$

Where t = the thickness of the pipe in inches.

„ H = the head of pressure in feet of water.

„ d = the inside diameter of the pipe in inches.

The following table for cast-iron pipes has been calculated from this formula for a head of 300 feet of water pressure :—

Internal diameter.	Length (not including socket).	Thickness of metal.		Weight (including socket).	Bursting pressure per sq. in.	Factor of safety for 300 ft. = 139 lbs. per sq. in.	Weight of lead joints.
		Inches and decimals.	Nearest thickness in 16ths of an inch.				
Ins.	Feet.			cwts.	lbs.		lbs.
2	6	·31	5/16	·418	4900	38	1·4
2½	6	·33	3/8	·625	3920	30	1·6
3	9	·35	3/8	1·06	3920	30	2·3
4	9	·375	3/8	1·38	2940	23	4·0
5	9	·41	7/16	2·01	2744	21	5·0
6	9	·45	7/16	2·38	2290	18	6·5
7	9	·47	1/2	3·17	2240	17	7·7
8	9	·50	1/2	3·59	1960	15	·8·2
9	9	·53	9/16	4·55	1960	15	10·4
10	9	·56	9/16	5·03	1764	14	11·5
11	9	·59	9/16	5·51	1604	12	13·5
12	9	·62	5/8	6·69	1633	13	18·0

As absolute correctness cannot for practical reasons be obtained, it is usual to allow a deviation of 3 per cent. in the calculated weights.

Cast-iron pipes are connected by means of flange-joints bolted together, by spigot and socket joints run solid with lead, etc., or the spigot is turned and the socket bored out to receive it. The flange-joint (Fig. 43) is stronger than either of the latter, but is more costly, and is rarely used when the pipes are laid horizontally in trenches, except for high pressures. It is chiefly used where the pipes are fixed

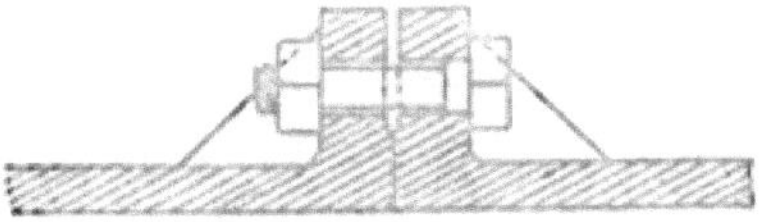

FIG. 43.

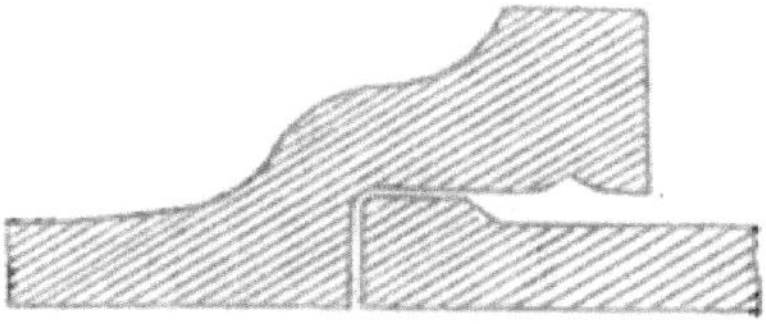

FIG. 44.

vertically for standpipes, etc. The faces are machined and jointed together with red-lead or with packing-rings made of lead, rubber, or other material.

The proportions of cast-iron flange-pipes are given in the following table:—

Diameter of pipes.	Diameter of flange.	Centre to centre of holes.	Thickness of flange.	No. of bolts in flange.	Diameter of bolts.
Inches.	Inches.	Inches.	Inch.		Inch.
2	6	$4\frac{3}{4}$	$\frac{3}{4}$	4	$\frac{1}{2}$
$2\frac{1}{2}$	$6\frac{1}{2}$	$5\frac{1}{8}$	$\frac{3}{4}$	4	$\frac{9}{16}$
3	7	$5\frac{3}{4}$	$\frac{13}{16}$	4	$\frac{9}{16}$
4	$8\frac{1}{2}$	$6\frac{3}{4}$	$\frac{7}{8}$	4	$\frac{9}{16}$
5	$9\frac{3}{4}$	8	$\frac{7}{8}$	6	$\frac{5}{8}$
6	11	$9\frac{1}{4}$	$\frac{7}{8}$	6	$\frac{5}{8}$
7	$12\frac{1}{2}$	$10\frac{1}{2}$	1	6	$\frac{5}{8}$
8	$13\frac{3}{4}$	$11\frac{1}{2}$	1	6	$\frac{3}{4}$
9	15	$12\frac{3}{4}$	1	8	$\frac{3}{4}$
10	16	$13\frac{3}{4}$	$1\frac{1}{16}$	8	$\frac{3}{4}$
11	$17\frac{1}{2}$	15	$1\frac{1}{8}$	8	$\frac{3}{4}$
12	$18\frac{1}{2}$	16	$1\frac{1}{8}$	10	$\frac{3}{4}$

Spigot and socket joints with lead, rust, rubber, or turned and bored joints take several forms. Fig. 44 is the joint adopted in the Tansa new works for supplying Bombay. Figs. 45A, 45B are frequently used, the former having the advantage of preventing blown joints, owing to the resistance offered by the bead on the spigot and the recess in the socket. The socket space for jointing should not exceed ¼ inch in thickness in pipes up to 3 inches diameter, $\frac{5}{16}$ inch from 3 to 8 inches diameter, and ⅜ inch from 8 to 12 inches diameter. Rust joints are rarely adopted for waterworks, and when used under special circumstances they do not differ from the forms shown for lead. They are made by forcing a mixture of iron borings or turnings and sal-ammoniac into the space between the socket and the spigot.

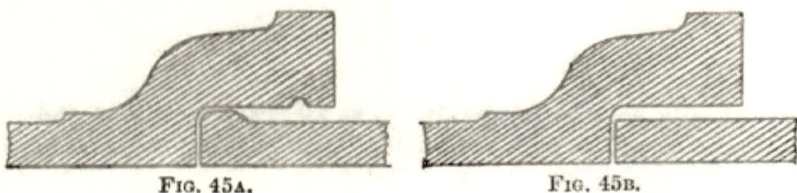

FIG. 45A.   FIG. 45B.

Rubber joints (Fig. 46), known as Forster's patent, are largely used in the north of England. Two beads are cast on the spigot end, between which one or more circular rubber rings are placed and then driven into the socket. Turned and bored joints (Fig. 47) have been extensively used in the north of England and abroad, but only in a few instances in the south of England. Pipes jointed thus must

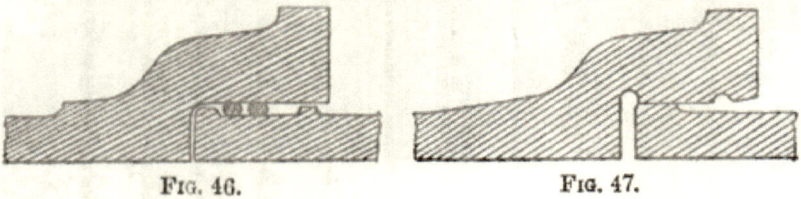

FIG. 46.   FIG. 47.

be laid in straight lines, with expansion joints or ordinary spigot and socket lead joints every tenth pipe, to allow for the variations in temperature. The half turned and bored is generally used, so that lead may be inserted if deemed necessary. The taper of the machined portion should not be more than 1 in 32, and the width should not exceed 1 inch. An increased width causes greater rigidity, rendering the work more liable to fracture by the traffic and superincumbent earth. Cast-iron pipes should be made from the grey variety, of good tough quality, which should be re-melted in a cupola before running. The increase in strength and density caused by re-melting is strikingly illustrated by the results recorded by Sir Frederick Bram-well with Acadian cold-blast iron, as follows: —

Samples.		Tensile strength per square inch.
1st samples	...	7·5 tons.
2nd ,, after 2 hours longer fusion ...	...	8·3 ,,
3rd ,, ,, 1¾ ,, ,, ...	...	10·8 ,, .
4th ,, re-melted with fresh pigs ...	...	11·0 ,,
5th ,, after 4 hours longer fusion ...	...	18·5 ,,
Maximum of 5th samples...	...	19·6 ,,

the tensile strength being increased 150 per cent. by eight hours of continued fusion. A clause should be inserted in a specification for cast-iron pipes, providing that a test-bar, cast from time to time, shall, when placed edgeways on bearings 36 inches apart, support a certain weight.

Test-bars 1 inch × 2 inches in section and 3 foot 6 inches long should be capable of supporting a weight of 30 cwts. gradually applied at the middle of the bar.

Pipes should be cast socket downwards in dry sand-moulds, and run, as quickly and equally as possible, in one operation, so as to avoid a "cold shut." Pipes 4 inches and upwards in diameter should be cast vertically; under 4 inch, at an angle of 45°. The strength at the spigot-end is increased by casting a head or additional 6 inches or more beyond the finished length of the pipe, which is afterwards cut off in the lathe. The head has the effect of compressing the metal, and permits the ash and bubbles to rise into and be removed with it. The pipes should be straight, cylindrical, and free from chaplets, core nails, and other imperfections, and the metal should be of uniform thickness throughout. The consecutive number, year, and maker's name should be cast on each pipe, the numbers on rejected pipes being disfigured by a chisel-cut, and no number of a rejected pipe must be replaced.

CHAPTER XVIII.

PIPES—*continued.*

THE pipes, after being cleaned, are struck all round with a light hammer, and, if sound, are placed in the testing machine (Fig. 48) and tested by oil or water; the former is perhaps better for the iron-work, but is more costly, and the benefits derived therefrom small. Gaskets or steel rings wrapped with yarn are hung at each end of the pipe, to form a joint with the iron plates, one of which (*a*) is fixed, and the other (*b*) movable, the latter being driven forward by a screw and

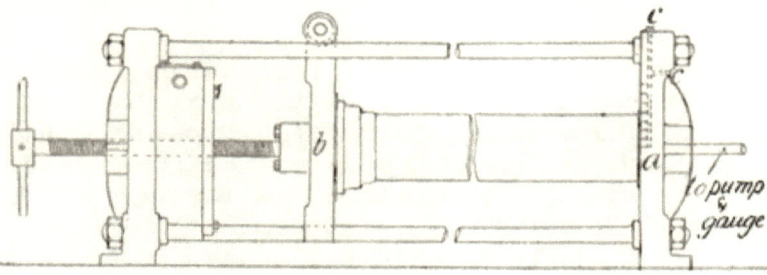

FIG. 48.

gearing worked by hand. The screw-plate has an air-pipe (*c*) cast in it, through which the air escapes, as the water, flowing in, fills the pipe under test. When the water flows full-bore through the air-pipe, showing that the air has escaped, the valve is closed and the screw-plate tightened up, the pump having been started just before. The pressure-gauge, which should be placed in a conspicuous position, records the rise of pressure. Attention must be given during the whole of the testing operation to the behaviour of the pipe. When the testing limit is reached, a valve between the pump and the testing-machine is closed, shutting off the connection with the pump and allowing the pipe to remain under the specified pressure while it is examined and struck with a hammer. During this period the gauge should remain stationary, unless there is a loss from leakage, which must at once be stopped and the test re-started. After being tested for pressure, the pipe is tested for thickness and uniformity of metal by means of calipers and a disc (Fig. 49). The examination being satisfactory, the pipe is conveyed to a heating-stove and raised to a temperature of 400° Fahr.,

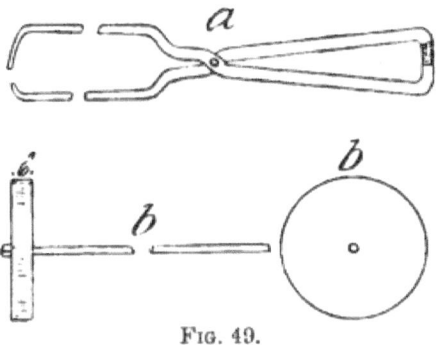

Fig. 49.

previous to dipping it in what is termed Dr. Angus Smith's composition, consisting of pitch, asphalt, resin, and linseed oil at a temperature of 300° Fahr. The pipe is then ready for delivery on the site of the works. The pipes are lowered into the trench prepared for them, either by hand by means of rope, or by a block-and-tackle arrangement attached to tripod legs. The trench is sufficiently enlarged at the junctions of the pipes, so as to admit of the jointing material being properly filled, caulked, and examined. Each pipe is struck with a hammer for soundness, and the spigot end carefully driven up into the socket of the preceding pipe, after which it is ranged in line and set at the required level, attention being paid to the even thickness of the joint. The

K

lead joint is made by driving a gasket of strip lead or a few coils of yarn into the space between the spigot and the socket by means of a yarning-iron (Fig. 50, *a*). The former method is coming into general use, and is preferable on sanitary grounds, as the yarn becomes a nest for bacteria. The gasket having been driven tightly up to the back of the socket, the joint is ready for the clay luting, which is placed around the lead space at the face of the socket, with a lip at the top to receive the molten lead. Care must be taken to remove the dross before running the molten lead, which must in all cases be done in one operation, so as to ensure a solid joint throughout. The lead having become solidified, the clay luting is removed, the surplus lead at the lip cut off, and the joint set up at least $\frac{1}{8}$ inch within the socket by means of caulking-tools (Fig. 50, *b*). After each joint is completed the pipe should be examined inside to see if any lead has run through from careless yarning, so that it may be removed. The joints of turned and bored pipes are made by painting the machined

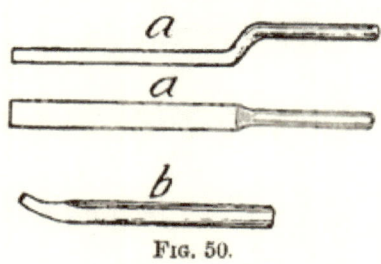

Fig. 50.

portion with thin red lead or liquid Portland Cement. The spigot is then placed within the socket and driven up with a wooden maul. The labour entailed in laying this class of pipes is small, and the general experience where they have been adopted is that they are quite equal to lead joints except for turning curves, where the lead joint may be utilized as well as for expansion purposes already referred to. Wrought-iron or steel pipes are largely used in the smaller sizes for connecting services in the place of lead pipes, where the water is of such a nature as to attack the lead and produce lead-poisoning. The pipes consist of strips of metal either welded or solid drawn, with screwed socket-joints; they are usually galvanized, and can be manufactured to suit any pressure required. Wrought-iron tubes are made in lengths usually not exceeding 14 feet, but can be obtained up to 20 feet in length if necessary. They are made in three qualities —gas, water, and steam—the steam-tubes being two gauges, and the water-tubes one gauge thicker than the gas-tubes.

The following table give the results of Messrs. J. Russel's experiments with wrought-iron solid-drawn tubes, according to Mr. D. K. Clark.

External diameter.	Thickness.		Internal diameter.	Bursting pressure.		Collapsing pressure.		Difference between bursting and collapsing pressure.
				Per sq. in. of surface.	Per sq. in. of section of metal.	Per sq. in. of section.	Per sq. in. of section of metal.	
Inches.	B.W.G.	Inches.	Inches.	lbs.	Tons.	lbs.	Tons.	Tons.
3¼	10	·134	2·982	4800	23·84	3300	17·86	5·98
3⅛	10	·134	2·857	4500	21·42	3150	16·40	5·02
3	11	·120	2·760	4500	23·10	3500	19·53	3·57
2¾	11	·120	2·510	5200	24·28	3500	17·89	6·39
2½	11	·120	2·260	5000	21·02	3600	16·74	4·28
2¼	11	·120	2·010	5900	22·06	4500	18·82	3·24
2	12	·109	1·782	5900	21·53	4900	20·07	1·46
1¾	12	·109	1·532	5600	17·57	4000	14·33	3·24

The greater tensile strength of wrought-iron and steel, and their lightness compared with cast-iron, give them great advantages over the last metal in such cases where weight and strength are the main objects, although for cheapness and convenience in casting, as well as the greater thickness for corrosion, it is doubtful whether wrought-iron or steel will replace cast-iron in the manufacture of pipes. The action of the Bradford Corporation, in adopting them in their new works, may lead to their more general use for large mains in somewhat inaccessible districts, where the question of weight is a serious one.

Lead pipes are of almost universal application for service connections and interior fittings, on account of the facility with which they can be bent to suit the irregularities of structure, and it is a matter of importance that they should be of good quality and of sufficient strength for the purpose. The following table gives the sizes of pipe usually specified for service connections :—

$\frac{1}{2}$ inch diameter   6 lbs. per yard.
$\frac{3}{4}$ inch   „   9 lbs.   „
1 inch   „   12 lbs.   „
$1\frac{1}{4}$ inch   „   16 lbs.   „

Any of these pipes would stand a pressure of 500 feet head of water. Several methods have been proposed for preventing the solvent action of some waters upon the lead, one of which is to line the interior of the pipe with block tin ; but none of the many proposals have been largely adopted.

Clay pipes are frequently used for conveying water in collecting-drains and other situations where there is no head of pressure on the pipes. The joints are either left dry or filled with Portland cement, as the circumstances require.

The following table of the dimensions of clay pipes was adopted by Mr. Baldwin Latham for the Bideford Waterworks :—

Internal diameter.	Stoneware.		Fireclay.		Other clays. Thickness.	All pipes, depth of socket.
	Thickness.	Length in work.	Thickness.	Length in work.		
Inches.	Inches.	Feet.	Inches.	Feet.	Inches.	Inches.
2	—		—		$\frac{5}{16}$	—
3	$\frac{3}{8}$	2	$\frac{3}{8}$	2	$\frac{3}{8}$	$1\frac{1}{2}$
4	$\frac{1}{2}$	2	$\frac{1}{2}$	2	$\frac{1}{2}$	$1\frac{1}{2}$
6	$\frac{5}{8}$	2	$\frac{5}{8}$	2	$\frac{5}{8}$	$1\frac{1}{2}$
9	$\frac{3}{4}$	2	$\frac{3}{4}$	2	$\frac{7}{8}$	2
10	$\frac{7}{8}$	2	1	2	—	—
12	$1\frac{1}{8}$	2	$1\frac{1}{16}$	2	1	2
15	$1\frac{1}{8}$	2	$1\frac{1}{4}$	2 to 3	$1\frac{1}{4}$	$2\frac{1}{4}$
18	$1\frac{1}{4}$	2 to 3	$1\frac{1}{2}$	2 to 3	$2\frac{1}{2}$	$2\frac{1}{2}$

When extensive works are being carried out it is customary to appoint an inspector to be present during the entire operations of casting and testing at the foundry. This entails a considerable expense which cannot be borne by a small undertaking. As the pipes are liable to suffer from shocks and concussions from rough handling and other causes in transit from the foundry to the site of operations, as well as from injuries in the course of jointing and lowering into the trenches, such inspection alone is not always satisfactory. In support of this view the experience recorded in connection with the waterworks construction at Market Harborough by Mr. H. G. Coates, A.M.I.C.E., may be instanced. The pipes were cast and tested at the foundry under the supervision of an inspector, the tests being conducted in accordance with the following schedule :—

Diameter of pipe.	Thickness of pipe.	Proof pressure equal to a column of water.	Or lbs. per sq. in.	Maximum pressure in feet to which pipe will be subjected in work.
Inches. 10	Inches. ½	Feet. 450	195	205
8	½	450	195	205
7	7/16	300	130	20
6	7/16	450	195	205
4	3/8	500	217	215
2	5/16	500	217	210

The pipes being struck with a hammer in all their parts whilst under pressure. Test-bars of the metal actually used were also carefully proved.

The whole of the system, comprising 8¼ miles of 10-inch, 65 yards of 8-inch, 2 miles 1410 yards of 6-inch, 5 miles 850 yards of 4-inch, and 1380 yards of 2-inch pipes, were

subsequnently tested in sections in the trenches by hydrostatic pressure to 50 lbs. per square inch above the working
pressure. The following is a list of the rejected pipes
resulting from this second test—

460 10-inch pipes
89 6-inch „
97 4-inch „
10 2-inch „

weighing about 141½ tons, or over 8 per cent. of the weight
delivered.

To enable the process of testing in the trenches to be
conducted economically, the pipes should be laid from the
reservoir towards the site of distribution, so that the water
may follow the work, and the cost saved of filling the
sections to be tested by water-carts.

The method of testing is as follows. The terminal pipe
is closed by means of a blank socket, drilled and tapped to
receive the connection-pipe from the pressure-pump. The
connecting-pipe may consist either of a short length of
strong hose—which is undoubtedly the most convenient—
or of wrought-iron tube with proper bends. In either case
the connection with the pump should be made by means of
a screw-union, so that the machine can be easily discon-
nected. The pump usually consists of a small pressure-
pump, similar to that used for testing boilers, mounted
over a tank or reservoir, which is frequently placed on
wheels. A water pressure-gauge, showing the pressure in
feet of water and in lbs. per square inch, is attached where
it can be easily seen, also a safety-valve, and valve for
lowering the pressure.

The section of pipe to be tested is then slowly charged
with water by opening the sluice-valves between it and
the reservoir, the air in the pipes being got rid of by
means of the air-valves and ball-hydrants (if any) on the
section. It is well to open the valve for lowering the
pressure, above referred to, and to allow the escaping
water to carry the air with it. The importance of getting
rid of as much air as possible is great, both on account of

the time and labour saved in getting up the pressure, as well as the danger from flying pieces of pipe, etc., were a burst to take place. In a small waterworks now being carried out the two following tests were made :—

1. Length of section, 420 yards ; diameter of pipes, 3 inches ; pressure attained, 230 lbs. per square inch.   Time, 10 hours.

2. Length of section, 250 yards ; diameter of pipes, 2 inches ; pressure attained, 230 lbs. per square inch.   Time, 3 minutes.

In the former case the line of pipes was undulating, and the air lodged in the high points.   In the latter case the pipes were laid at a uniform slope.   Having dislodged as much air as possible, the valve for lowering the pressure and the nearest sluice-valve should be tightly closed and pumping commenced.   The pump-tank is kept constantly filled by means of a water-cart or by buckets, the quantity of water required depending upon the amount of air in the pipes and the number and extent of leakages.

The line of pipes should be carefully examined while the pressure is rising, and pegs driven or marks made at any sign of failure.   When the pressure has risen, as shown by the gauge, to the proof point, which should be at least double the working head, it should be kept there while the pipes are submitted to a close scrutiny.   It must be noted if the needle remains steady when pumping ceases, or falls more or less rapidly.   The behaviour of the needle under these circumstances greatly depends upon whether the instrument is situated at the highest or lowest level of the section under test.   In the former case a slight leak will cause the needle to fall **rapidly** ; in the latter the needle will be scarcely affected.

No part of the trench must of course have been filled in until the testing has been satisfactorily completed, and any water in the trenches must, if possible, be removed, otherwise the examination is bound to be imperfect.

The **most** familiar failures are split or cracked pipes, honeycombed-sockets, pin-holes, blown core-nails, and leaking joints. Each point of failure must be carefully recorded by means of a peg or intelligible mark. The pressure is then lowered, defects made good, and the section re-tested.

Pin-holes in otherwise sound pipes may be drilled, tapped, and plugged with brass or gun-metal plugs, but only with the special permission of the engineer in each case. Weeping joints may frequently be remedied by the application of the caulking-tool. Honeycombed or perforated sockets, or split pipes, must be cut out, and the line of pipe made good by means of a short piece cut to the requisite length, and a thimble or sleeve.

Corrosion or rusting of iron pipes, both externally and internally, is an element not to be lost sight of by the engineer. Externally it is caused either by the oxygen in the water attacking them where the pipes are laid in damp ground, or by the presence of corrosive substances in the soil, which occasionally occurs in the neighbourhood of chemical works, etc. The former, in the case of cast-iron pipes, may be neglected if they are properly protected by coating with Dr. Angus Smith's solution, on account of the comparatively great thickness of the shell; the latter must be prevented, where necessary, by embedding the pipes in concrete. Internally, corrosion is caused by the oxygen in the water attacking the metal, and is usually more serious with soft than with hard water. The coating of the interior of the pipes with the protecting solution must be carefully attended to, and made a special point in their inspection. Even with the most careful coating corrosion is almost certain to occur, but the process will greatly delay it; minute points are left unprotected, and corrosion will take place, gradually scaling off the intervening varnish where these points are close together. The danger of weakening the pipes is small, for the reason above stated; but the real trouble lies in the contraction of the bore of the pipes caused by the nodules of rust (the rust does not form in even layers), which diminishes their discharging capacity and reduces the working head. Mr. Thomas Box ("Practical Hydraulics") records a case at Torquay where a main, about 14 miles long, composed of 14,267 yards of 10-inch, 10,085 yards of 9-inch, and 170 yards of 8-inch pipe delivered only 317 gallons per minute, with 465 feet head, or only 50 per cent. of the theoretical discharge. The pipes were subjected to repeated scraping by means of scrapers worked through their entire length, the result of which was that the experimental discharge was brought up to the theoretical.

On account of corrosion, wrought-iron tubes for conveying water should always be galvanized. In designing systems of pipes it is always advisable to increase the calculated diameter so as to allow for corrosion.

The following memoranda will be useful to the student when calculating the weight of pipes :—

Wrought-iron, 1 cubic inch, 0·278 lbs.; 1 cubic foot, 480 lbs.

Cast-iron, 1 cubic inch, 0·260 lbs. ; 1 cubic foot, 450 lbs.

Steel, 1 cubic inch, 0·283 lbs.; 1 cubic foot, 489·6 lbs.

Lead, 1 cubic inch, 0·412 lbs. ; 1 cubic foot, 712 lbs.

Gun-metal, 1 cubic inch, 0·304 lbs. ; 1 cubic foot, 524 lbs.

CHAPTER XIX.

FIRE-SERVICE, VALVES, AND METERS.

THE details necessary for the provision of an efficient fire-service too often receive but scanty attention in waterworks construction, especially in rural districts. The advantages of having a powerful stream of water, easily put into requisition, and capable of playing upon any portion of a building in a state of conflagration, are too great to need impressing upon the student. It is necessary, however, to point out that these facilities can frequently be attained at very little extra cost, if the necessary arrangements are included in the original design of the waterworks, whilst the cost of their subsequent provision may be prohibitive. The principal points to be kept in view in making provision for fire-service are :—

1. That there shall be a surplus storage of water for fire extinction over and above that required for general purposes, and always available.

2. Such storage to be at a sufficient elevation to allow of the water being forced above the tops of the highest buildings in the district.

3. The mains and distributing pipes to be of such dimensions as to allow of the water for fire extinction purposes being conveyed through them when the demand for water for other purposes is at its greatest.

These requirements cannot always be secured in their entirety, but they indicate the lines which must be kept in view. The surplus storage is usually included in the capacity of the service-reservoir, and the amount to be allowed for must depend upon the special requirements of the district under consideration, as well as upon the means at the disposal of the engineer.

Assume, for the sake of example, that on the grounds of probability and expediency, only one fire at the same time in the district is to be provided for, that the probable time occupied in extinguishing the fire will be three hours, and that one jet, 50 feet high, will be required. Theoretically, the height of the water issuing vertically from a jet, should be equal to the head upon the jet, but the resistance of the air causes a considerable reduction. Experiments show that the difference between the theoretical height and the actual height attained by a jet of water varies approximately, directly as the square of the theoretical height, and inversely as the diameter of the jet.

Mr. Box (" Practical Hydraulics") gives the following formula :—

$$h' = \frac{H^2}{d} \times \cdot 0125$$

Where H = the head on the jet in feet.

    „ h' = the difference between the height of the head and the height of the jet.

    „ d = diameter of the jet in $\frac{1}{8}$ths of an inch.

It is evident from the above relations that in order to obtain the best results from the available head, a jet of special diameter corresponding to that particular head must be used. Assuming an available head of 70 feet at the point under consideration, then—

$$h' = 70 - 50$$
$$d = \frac{H^2 \times \cdot 0125}{20} = 3$$

or the proper diameter of jet to be used, under the circumstances, is $\frac{3}{8}$ths of an inch. The quantity of water discharged by a jet with a given head depends upon the form of the jet (see Chap. X., gauging by means of an orifice), the best form being that which approaches most nearly to the *vena contracta* (Fig. 51). The discharge through this form of jet may be taken as ·943 of the theoretical discharge due to the head.

Applying this coefficient to the formula—

$$Q = 8 \cdot 025 \, c \, a \sqrt{h}$$

$$Q = 8 \cdot 025 \times \cdot 943 \times [(\tfrac{3}{8})^2 \div 144 \times \cdot 7854] \times \sqrt{70}$$
$$= \cdot 04856 \text{ cubic feet per second.}$$
$$= 18 \cdot 21 \text{ gallons per minute.}$$

If the jet is required to play for 3 hours, then the quantity of water discharged will be $18 \cdot 21 \times 60 \times 3 = 3277 \cdot 8$ gallons. It would, therefore, be necessary to provide surplus storage capacity in the service-reservoir for, say, 4000

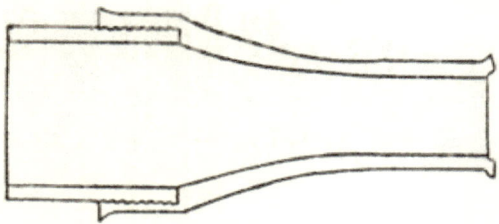

Fig. 51.

gallons; and the diameter of the mains and distributing pipes must be chosen so as to allow of an extra quantity, equal to 18·21 gallons per minute, being conveyed through them when the demand for water for other purposes is at its greatest.

In order to render a fire-service efficient, it is necessary that a sufficient number of hydrants should be provided upon the system, the situations being selected with great care, so that each may yield the maximum of efficiency. They should be easily accessible, capable of rapid manipulation, not difficult to find in the dark, and so constructed as not to be affected by frost.

In designing a waterworks system, the available head at any point is arrived at by deducting the loss of head due to friction in the pipes from the statical head at that point (Chap. XI.). It must be remembered that these results will only hold good provided the system is watertight, allowing practically no waste. Sir Frederick Bramwell, speaking of the evils of waste, at a recent meeting of the Institution of Civil Engineers, expounded this point very clearly.

"Another point was the diminution of the pressure in the pipes. It was impossible to obtain an extra amount of water through a given-sized main, except by having a greater differential pressure between the entering end and the delivery end of the main. The entering end of the main was fixed by a reservoir or a stand-pipe level, or whatever it might be, and the consequence was that the increase in the differential pressure must be obtained by diminishing the pressure at the delivery end. What was the result? Any hope of using the water for hydrant purposes, for extinguishing fires, was gone: there was no pressure for that."

The two principal types of hydrants or fire-cocks are the sluice-valve and the ball-hydrant. There are innumerable forms in the market, but they are all modifications or combinations of these two types. The sluice-valve hydrant consists, as its name implies, of a sluice-valve connected with the main, with a bend attached to its outlet fitted at its upper end to receive a stand-pipe to which the hose is fixed. The stand-pipe is either fitted to the hydrant by a screw or bayonet-joint. The stand-pipe is sometimes a fixed pillar, which has the advantage of easier access at times of fire. The sluice-valve is the best form of hydrant, but entails a heavier first expense than the ball-hydrant. A frost-cock, which may be automatic, must be attached on the outlet side of the sluice-valves, for the purpose of removing the water which would otherwise remain in the bend (or pillar) after use.

The ball-hydrant, patented by Messrs. Bateman and Moore, consists of a vulcanite ball contained in a valve-box; the outlet to the box, which is vertically above the ball, and is fitted with a leather or indiarubber washer, being kept constantly closed while the water in the main is under pressure, by the ball, which is lighter than the water, being forced up against it. The stand-pipe is attached to the hydrant by means of a bayonet-joint; and the valve is opened by depressing the ball by means of a spindle passing down through the stand-pipe and worked by a crutch-handle. Ball-hydrants are economical, and work exceedingly well with moderate heads. With low heads they are apt to leak, and with high heads the ball is liable to be forced out of shape, causing leakage when it takes a new bearing upon its seat. Another objection is caused by the suction into the main through the ball-hydrant, when the former is being emptied of any liquid matter, frequently of a filthy nature, which may be at the time in the hydrant-chamber. Ball-hydrants also act as air-valves, but with very doubtful advantage, on account of the large orifices, which allow the air to escape so rapidly, while the main is being charged, as to endanger the pipes by shocks.

Air-valves are used for the purpose of getting rid of the air which constantly accumulates at the highest points of undulations in the line of pipe, especially where such points are situated above the hydraulic mean gradient (see Chap. XI.). They are either automatic, in which case they are identical with the ball-hydrant, except that the aperture which serves for the escape of the air is much smaller ($\frac{1}{8}$ to $\frac{3}{8}$ inch in diameter); or they consist of small stop-cocks, opened and closed by hand.

Sluice-valves vary little in form, and usually consist of a cast-iron body containing a movable diaphragm which slides vertically between grooves. The sliding faces, both of the body and of the movable valve, should be made of gun-metal, as well as the screw which actuates the valve and the stuffing-box gland through which the screw works. It is best to obtain them with spigot and socket ends attached by means of bolts and nuts, so that they can be removed if necessary without cutting the pipes. Sluice-valves should be fixed at all branches in a waterworks system; and the mains and branches should be divided up into easily worked sections by means of sluice-valves, so that each section may be isolated from the rest with a minimum of inconvenience to the consumers generally. Sluice-valves should be plentiful on all waterworks systems, and it is false economy to attempt to make a saving by reducing their number.

Water-meters are inserted in a line of pipe for the purpose of measuring and registering the flow of water passing through them. There are two types—the positive and the inferential. The positive meter measures the flow of water by causing it to alternately fill and empty a vessel of known capacity, the number of times that this process takes place being recorded by means of a clockwork mechanism. The positive meter is sub-divided into high and low-pressure meters—the Duncan, Kennedy, Frost, Schönheyder, Frager, and Kent meters representing the former; the Parkinson and Tylor representing the latter class. The inferential meter consists of a chamber through which the water flows, containing a wheel with vanes or discs attached. The water in passing impinges upon the vanes and causes the wheel to revolve, the revolutions being recorded as in the positive meter. The Siemens, Tylor, and Sporton meters are instances of this type. The mechanism of the positive meter is similar to that of the cylinder and slide-valve of a high-pressure steam-engine, while that of the inferential meter may be compared to a water-wheel or turbine. Messrs. Turner and Brightmore ("The Principles of Waterworks Engineering") give the following essentials as characteristic of a perfect water-meter :—

1. Accurate registration of the quantity of water passing through it, whether great or small.

2. Ability to perform its work without causing a material loss of head in the supply-pipe.

3. Cheapness and simplicity.

4. Ease of attachment and repairs.

5. Freedom from excessive wear of the working parts.

Section 58 of the Public Health Act, 1875, empowers a Local Authority "to agree with any person to supply water by measure, and as to the payment to be made in the form of rent or otherwise for every meter provided by them." The question as to whether the supply of water for domestic use should be charged for by measure has been largely discussed. There is a strong feeling against this method on sanitary grounds, so far as it applies to houses of low rental ; the objection raised against it being that water would be economized at the expense of cleanliness and health, in the very situations where these are of the greatest importance to the general community. Where the water is supplied for trade or manufacturing purposes the case is different, and, where it can be arranged, the sale should be by meter. This question is of the most importance where the water has to be pumped.

Meters have not been largely introduced into rural districts, although they can be applied with great benefit to farm—especially dairy farm—supplies. In a dairy farm where the milk is refrigerated, the consumption of water for that purpose alone is frequently 1000 gallons a day.

As the size of the pipe for conveying a supply of water to any premises frequently depends upon other considerations than its capacity for delivering that supply, it must not be taken as the gauge of the meter to be inserted on its line. The following table, which refers to Sieman's inferential meter, will be found useful in deciding upon the size of meter applicable in any particular case :—

No.	Inch.	Delivery in gallons per hour.	
		50 feet head.	150 feet head.
1	$\frac{3}{8}$	150	250
2	$\frac{1}{2}$	300	500
3	$\frac{3}{4}$	600	1,000
4	1	1,500	2,500
5	$1\frac{1}{4}$	2,200	3,800
6	$1\frac{1}{2}$	3,000	5,000
7	2	4,000	7,000
$7\frac{1}{2}$	$2\frac{1}{2}$	6,000	10,000
8	3	8,300	14,000
9	4	13,400	23,000
10	5	18,500	32,000
11	6	27,000	46,000
12	8	45,000	77,000
13	10	70,000	120,000
14	12	90,000	154,000

It is always better to allow a safe margin in deciding upon the size of a meter. Where the water is expensive, either on account of pumping or where it is purchased by measure by the undertakers in the first instance, it will be found a wise precaution to insert meters on all large connections as a check upon waste and undue consumption, where the consumer is not to be charged by measure.

CHAPTER XX.

HOUSE CONNECTIONS AND FITTINGS.

In order to secure satisfactory results in a waterworks undertaking, it is necessary to keep a strict control over the connections that are made by consumers with the undertakers' pipes for the purpose of obtaining a supply of water for domestic or other purposes. Waste is the *bête noire* of the waterworks engineer, and experience has shown that its favourite lair is to be found in the communication-pipes, taps, cisterns, and overflow-pipes connected with house and trade supplies.

As indicated in Chapter XIX., the difficulty where it applies to trade connections can be satisfactorily met by the insertion of water-meters, and by charging for the actual quantity of water delivered to the premises in question. When persons have to pay for water by meter, whether used or wasted, it becomes to their interest to detect the sources of waste; on the other hand, it is against the immediate interest of owners of house property to detect leakages or faulty fittings, which they would have to repair or replace at their own cost, without obtaining thereby any reduction in their water-rates.

To enable a Local Authority, *inter alia*, to suppress waste, the Waterworks Clauses Act, 1863, and certain provisions of the Waterworks Clauses Act, 1847, were incorporated with the Public Health Act, 1875. These enactments afford very meagre powers to Local Authorities, so far as giving them any control over domestic services is concerned. It is customary for the undertakers, when seeking for a special Act, to apply for extended powers. These powers, when obtained, are set forth in the form of " water regulations," to be considered in the next chapter. It may here be observed, parenthetically, that no provision is made in the Public Health Acts, empowering Local Authorities to make water regulations.

The sections of the Waterworks Clauses Acts (1863, and part of 1847) directly or indirectly enabling Local Authorities to place some check upon the waste of water in domestic services are the following :—

The Waterworks Clauses Act, 1847.

Sec. 28 empowers the undertakers to break up streets, etc., for the purpose of supplying water to the inhabitants of the district.

Sec. 44 requires the undertakers to lay down communication-pipes and other necessary works to any dwelling-house (under £10 rental) situated in any street where they have laid pipes (1) either at the request of the owner, or (2) at the request of the occupier, and upon payment or tender of the proportion of water-rate in respect of such house by this or the special Act, made payable in advance. Such reasonable annual rent for such pipes to be charged as may be agreed upon, or as may be settled by two justices.

Under these circumstances, the undertakers are enabled to employ efficient workmen and fix proper fittings.

Sec. 48 empowers the owner or occupier, having paid or tendered to the undertakers the portion of the water-rate as indicated in sec. 44, to open the ground and lay leaden or other communication-pipes between his premises and the undertakers' pipes, provided—

1. That he has obtained the consent of the owner and occupier of the intervening ground ;

2. That the pipes are of a strength and material to be approved by the undertakers ;

3. That fourteen days' notice has been given to the undertakers.

Sec. 49 requires the owner or occupier to give two days' notice of day and hour when communication is intended to be made. Communication to be made under superintendence, and according to the directions of the undertakers' surveyor, or other officer appointed, unless such surveyor or other officer fail to attend at the time mentioned in the notice.

Sec. 50 enacts that the bore of such communication-pipe shall not exceed the prescribed limits; or, if no limit has been prescribed, ½ inch, except with the consent of the undertakers.

Sec. 52 gives owner or occupier the same privileges as the undertakers as to breaking up roads.

It will be evident to the student, after a perusal of these sections, that the undertakers' control over a domestic supply is limited to the communication-pipe, and does not extend to the fittings.

Sec. 54 applies only to intermittent systems, and enables the undertakers to require the consumer to provide a proper cistern, ball, and stop-cock, and to keep them in repair. It has been held that the undertakers cannot enforce the provision of a valve instead of a plug-cock.

Sec. 56 (in continuation of sec. 54) empowers the undertakers to repair such cistern, and to recover the expense.

Sec. 57 gives power of entry to the surveyor or other person acting under the authority of the undertakers, for the purpose of detecting waste or misuse of water.

Secs. 58 and 59 impose penalties on misappropriation of the water; and

Sec. 60 imposes a penalty upon any person who shall wilfully or carelessly break, injure, or open any lock, cock, valve, etc., the property of the undertakers.

The Waterworks Clauses Act, 1863.

Sec. 12. A supply of water for domestic purposes shall not include a supply of water for cattle, or for horses, or for washing carriages where such horses or carriages are kept for sale or hire by a common carrier, or a supply for any trade, manufacture, or business, or for watering gardens, or for fountains, or for any ornamental purpose.

Water used for watering a horse and washing a carriage has, however, been held to be used for " domestic purposes" within the meaning of an enactment similar to the above section. Water used for a pleasure garden was held to be for a " domestic purpose."

Sec. 14 gives the undertakers power to let meters, cisterns, pipes, etc., on hire.

Sec. 16 empowers the undertakers to cut off the supply, if any of the provisions of the Act are contravened.

Sec. 17 imposes a penalty upon the consumer for waste of water through non-repair of pipes, taps, etc.

Sec. 18 imposes penalty upon misappropriation of the water.

Sec. 19 imposes penalty upon any alteration to service as follows : "It shall not be lawful for the owner or occupier of any premises supplied with water by the undertakers, or any other person, to affix, or cause or permit to be affixed, any pipe or apparatus to a pipe belonging to the undertakers, or to a communication or service-pipe belonging to or used by such owner, occupier, consumer, or other person, or to make any alteration in any such communication or service-pipe, or in any apparatus connected therewith, without the consent, in every such case, of the undertakers."

Sec. 20 imposes a penalty upon any person using the undertakers' water without agreement.

The powers possessed by a Local Authority for the prevention or suppression of waste in domestic connections may be summarized as follows :—

1. Where the dwelling-house is under £10 rental, and a request has been made by the owner or occupier for a supply of water, the Local Authority have control over workmanship, materials, and choice of fittings.

2. Where the owner or occupier makes his own connection, the Local Authority may superintend the communication or junction with their pipes; in such case the communication

to be made under their direction. The communication-pipes to be of a strength and material to be approved by the undertakers, the bore not to exceed the prescribed limit, or, if no limit, ½-inch.

3. The Local Authority has power to enter any house supplied by them with water, between the hours of 9 a.m. and 4 p.m., for the purpose of detecting waste or misuse of water.

4. Where the supply is intermittent, the Local Authority may require the provision of a proper cistern to hold the water so supplied, with a ball and stop-cock in the pipe bringing the water from the works of the Local Authority to such cistern; also to require that such cistern, etc., shall be kept in proper repair so as to prevent waste; also to, themselves, repair such cistern, etc., and recover the expense.

5. Power to impose a penalty for waste of water through non-repair of pipes, etc.

6. Power to impose a penalty for any alteration to service without the consent, in each case, of the Local Authority.

7. Power to cut off the water if any of the provisions of the Act are contravened.

These powers are utterly inadequate to the proper control over a waterworks, and as it is unusual for a Local Authority (especially a Rural Authority) to go to the expense of a special Act, so-called "Water Regulations," which, however, cannot be legally enforced, are not infrequently formally passed and published by the Local Authority, and generally attain the required result.

To meet the difficulty of not having sufficient control over domestic fittings, a system of testing and stamping was introduced in 1883 by Mr. Ernest Collins, M.I.C.E., the engineer to the New River Company, London. Mr. Collins says, "The system then introduced has developed extensively; and there has been a material improvement in the quality of the fittings. Manufacturers who were at first antagonistic to the arrangements, have become strong supporters of the system, insomuch that the

use of untested and unstamped fittings is, in the district of
the company mentioned, almost the exception; and where
such fittings are used, they are invariably of the same
strength and proportions, and in accordance with the regulations adopted by the company."

The same system had been previously (1873) introduced
at Liverpool, with the same object, by the engineer, Mr. G.
F. Deacon, M.I.C.E. Of this system, Mr. Deacon says,
"Very little difficulty was experienced. Naturally the
plumbers at first rebelled, but in a short time they were
glad to have their names put on the backs of the waste-
water notices."

Where new connections are to be made with the undertakers' pipes for domestic or other supply, it is now almost
universally the practice for the undertakers to tap the
main themselves, and lay, at their own expense, the communication-pipe to the fence or frontage wall (if there be
one) forming the boundary of the street or highway in
which their main is situated. At the termination of the
pipe a stop-cock is fixed, to which the consumer attaches his
own work. There has been much discussion as to the
advisability or otherwise of the practice of fixing outside
stop-cocks, but in a rural system, at any rate, there can be
little doubt as to its usefulness. The principal advantages
are (1) Facility with which water can be turned off by the
undertakers in the case of waste, or when premises are un-
occupied; (2) Facility afforded for detecting waste in mains
or connections. This stop-cock, and all other cocks or taps
used throughout the premises, should be of the screw-down
type. The following are the requirements of the New River
Company: "All taps must be fitted with loose valves; and
such valves must be lifted by the spindle, and must not be
dependent upon the pressure of water for opening. They
must be fitted with washers of oil-dressed leather, and for
hot water with vegetable fibre of the best quality. Stop-
taps must have set screws to secure flanges. The word
'inlet' must be distinctly marked on the inlet side of the
tap. They must be made with screwed ends and unions.

Spindles must, in all cases, be of gun-metal. All other parts may be of brass, of good suitable quality. Screw-down fittings must have four threads of spindle in the cover when closed " (Fig. 52).

It is essential that all overflow pipes from cisterns, baths, etc., be constructed as warning-pipes, the mouth of such warning-pipe to be conspicuously placed so that any waste cannot fail to draw immediate attention. On no account must any overflow be allowed to escape directly into any waste-pipe or drain. An important point is to see that all

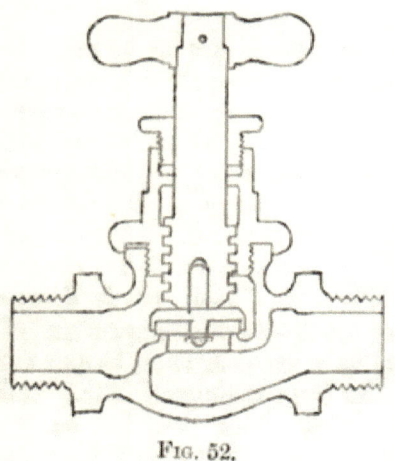

Fig. 52.

pipes are properly protected from the action of frost. This is rarely attended to, and consequently the waste during the winter months from burst pipes and taps, as well as the unnecessary expenditure of water caused by leaving taps running to prevent what should have been rendered impossible, is often enormous.

Instead of making a separate connection for each house, where the houses are close together and are of small rental, it is usually the custom for the undertakers to erect stand-posts or pillars from which the inhabitants of the houses can obtain their supply.

To facilitate this practice sec. 9 of the Public Health (Water) Act, 1878, enacts as follows:—

"Where a Rural Sanitary Authority have provided a stand-pipe for the supply of water to any portion of their district, they may recover water-rates or water-rents from the owner or occupier of every dwelling-house within 200

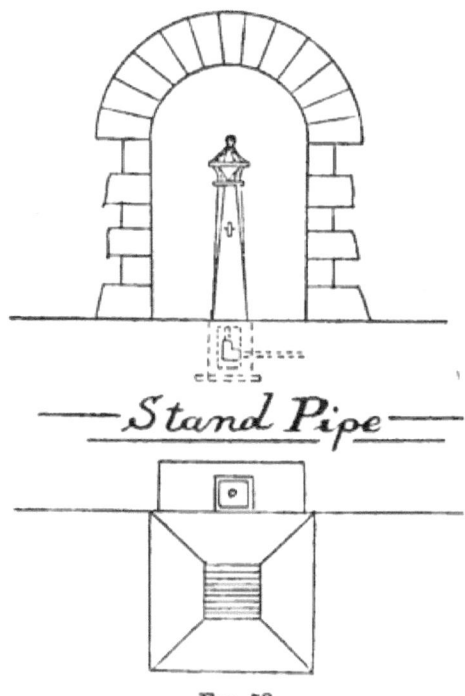

Stand Pipe

Fig. 53.

feet of any such stand-pipe, in the same manner in all respects as if the supply had been given on the premises.

"Provided that if any such dwelling-house has, within a reasonable distance, and from other sources, a supply of wholesome water sufficient for the consumption and use of the inmates of the house, no water-rate or water-rent shall be recoverable from the owner or occupier of the house unless and until the water supplied by means of such stand-pipes is used by the inmates of the house."

Stand-pipes consist of a $\frac{1}{2}$-inch or $\frac{3}{4}$-inch pipe running up a post or convenient wall, terminating in a tap fixed at such a height as will easily allow of a pail or bucket being placed under it when being filled (Fig. 53). The pipe is protected from the frost by means of a wooden or iron casing—the space between the casing and the pipe being filled with saw-dust or other non-conducting substance. The stand-pipe frequently consists of a strong cast-iron hollow pillar, the foot of which is firmly bedded into the

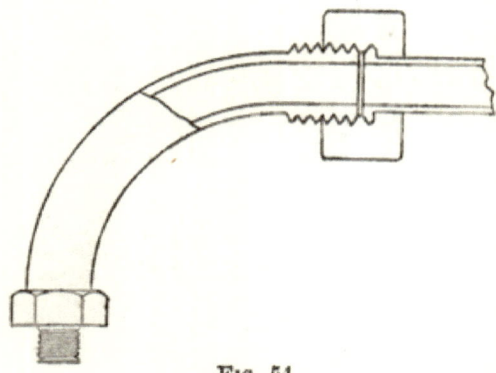

Fig. 54.

ground, the water-pipe passing up the centre. The tap for drawing off the water may be either a screw-down valve-cock or a self-closing cock. The latter is sometimes unsatisfactory with very high pressures. A stop-cock should always be inserted between the main and the stand-pipe to cut off the supply when the latter is being repaired. A grating should be placed under the tap, with a drain to carry away the waste water. A small tap is frequently placed at the foot of the stand-pipe to act as a frost-cock; the stop-cock is closed and the water in the stand-pipe emptied by allowing the water to flow away into the drain by means of the frost-cock.

Small connections are generally made with the mains by means of ferrules, which consist of small brass elbows, one end of which is screwed (Fig. 54) into a hole drilled and tapped into the top of the main; the other end has a union for attaching to the lead or wrought-iron service-pipe.

CHAPTER XXI.

WATERWORKS REGULATIONS.

In accordance with the provisions of the Metropolis Water Act, 1871, a code of regulations was compiled by the Metropolitan Waterworks Companies and submitted to the Board of Trade.

These suggested regulations were subjected to an exhaustive inquiry, and, as finally settled, have been circulated by the Local Government Board for the information of Local Authorities who have obtained the necessary powers, and are preparing to submit regulations for confirmation. As it is customary for Local Authorities supplying water, and acting only under the provisions of the Public Health Acts, to issue a code of Waterworks Regulations, trusting to their moral suasion and to the popular ignorance of the law for their efficacy (a trust by no means unfounded), a brief commentary upon the regulations under the Metropolis Water Act, 1871, is given here. These regulations are by no means perfect, or are they fitted for all cases. In adapting them for use in rural districts, latitude must be given to many of the provisions, care being taken that the main principles involved are not lost sight of.

No. 1 gives the company power to determine the point at which the communication-pipe shall enter the premises to be supplied.

No 2 requires that all lead pipes in direct communication with the company's system shall be uniform in thickness, and fixes the strengths—the weight being taken as the guide.

Internal diameter in inches.	Weight in lbs.		Thickness in inches.	Safe head of water in feet.
	Per yard.	Per foot.		
$\frac{3}{8} = $ ·375	5	1·67	·1002 $= \frac{3}{16}$	622
$\frac{1}{2} = $ ·50	6	2·0	·1873 $= \frac{3}{16}$	460
$\frac{5}{8} = $ ·625	7$\frac{1}{2}$	2·5	·196 $= \frac{3}{16}$	385
$\frac{3}{4} = $ ·75	9	3·0	·202 $= \frac{3}{16}$	330
1 $=$ 1·0	12	4·0	·212 $= \frac{3}{16}$	260
1$\frac{1}{4} = $ 1·25	16	5 3	·230 $= \frac{1}{4}$	226

The weight is calculated on a basis of 712 lbs. to the cubic foot, and the strength upon a mean ultimate cohesion of 2000 lbs. per square inch, allowing a factor of safety of 7$\frac{1}{2}$.

The weights adopted by various water companies, as might be expected, vary considerably, on account of the different pressures to which the pipes will be subjected.

The following are the weights, per lineal yard in lbs., required by various companies:—

Name of company.	Internal diameter of pipe in inches.					
	$\frac{3}{8}$ in.	$\frac{1}{2}$ in.	$\frac{5}{8}$ in.	$\frac{3}{4}$ in.	1 in.	1$\frac{1}{4}$ in.
London Companies ...	5	6	7$\frac{1}{2}$	9	12	16
Kent	—	5	7	9	12	—
West Surrey	4	5$\frac{1}{2}$	—	9	14	20
Caterham...	5	6	8	10	14	—
Colne Valley	5	7	9	11	16	—
Sevenoaks and Tonbridge	—	5	7	9	12	15
Norwich	5	7	9	11	16	22$\frac{1}{2}$
Sheffield	5	7	9	11	16	22$\frac{1}{2}$
Market Harborough ...	5	6	7$\frac{1}{2}$	11	16	20
Glasgow (Loch Katrine)	—	7	—	10	14	18

No. 3 allows the consumer the option of lead, copper, or

wrought-iron, for internal pipes, except when in contact
with the ground, when the company may insist on lead
being used.

No. 4 limits the consumer to one communication-pipe.
This is an important regulation.

No. 5 requires each house supplied by the company to
have its own communication-pipe, except in the case of a
block of buildings belonging to one owner, who pays the
water-rates for them.

It frequently happens in rural districts that the undertakers are asked to allow a communication-pipe to a property
to be tapped for the benefit of another property (the respective owners having agreed between themselves). This
should never be permitted.

No. 6 prohibits any communication between the pipes or
fittings of any two premises, except in the case provided for
in the last regulation.

No. 7 provides that the connection of the communication-
pipe with the company's pipes shall be made by the company at the cost of the consumer, the connection to be made
by means of a " sound and suitable brass-screwed ferrule or
stop-cock with union," having a clear waterway of not less
than that of a ½-inch pipe.

No. 8 requires that every pipe external to the house,
including the portions of pipes laid in external walls,
shall be of lead; the joints to be " plumbing " or " wiped "
joints.

No. 9 guards against possible pollution to the water in
the consumer's pipes, and thence to the water in the
company's pipes, from the consumer's pipes being " laid
or fixed through, in, or into " any drain, ashpit, etc. Where
such drain, ashpit, etc., is in the unavoidable course of the
pipe, the pipe must be protected by passing it through a
cast-iron pipe or jacket of sufficient length and strength
and of proper construction. When the water is turned off
from any part of the system a partial vacuum is formed,
and any liquid matter in proximity to a faulty pipe or
leaking joint would be sucked into the pipe, and might

be the cause of dissemination of disease throughout the whole system.

No. 10 requires that all pipes laid in open ground shall be laid at a depth of at least 2 feet 6 inches below the surface, and provides for proper protection against frost in exposed situations. A depth of 2 feet is frequently inserted in waterworks regulations, which is probably ample in this country.

No. 11 prohibits any communication between the pipes and any receptacles for rain-water.

No. 12 provides for the insertion of a " sound and suitable screw-down stop-valve," either at or near the point of entrance of the communication-pipe into the premises, or within the premises, at the option of the consumer. If placed in the ground, such stop-valve to be protected with a proper cover and guard-box.

For the reasons stated in Chap. XX., it is better to have the stop-cock fixed outside, so as to be under the immediate control of the undertakers. The principal objection, from the consumer's point of view, is that it is less accessible in case of burst pipes or other accidents, and entails the use of a loose key, which is liable to be mislaid.

No. 13 deals with cisterns, requiring that they shall be made water-tight, properly covered, placed in such a position that they can be easily inspected and cleansed, and that each cistern shall be provided with a sound and suitable " ball-tap " of the valve kind.

It is evident that if a cistern which is filled automatically is unsound, the waste of water must be constant. The provisions as to covering and cleansing allude to the possibility of pollution in a similar manner to that explained in connection with regulation No. 9.

No. 14 prohibits the use of overflow or waste-pipes other than " warning-pipes," to cisterns; and

No. 15 requires that all " warning-pipes " shall discharge at such a point that any flow may be readily ascertained by the officers of the company. The position of such "warning-pipes" not to be altered without due notice to and the approval of the company.

These regulations are of the greatest importance, and should be strictly enforced. The old practice of allowing overflow pipes to empty directly into the drains was a prolific source of waste. In the case of the ball-tap being out of order, the waste might continue for months without being detected. With a "warning-pipe," which consists of a short pipe passing directly through the wall into the air, with an open mouth, the case is different; any waste is speedily detected, and the inconvenience caused by a constant stream of water flowing down the face of the wall of the house, causes the occupier to take prompt steps to remedy the defect.

No. 16 prohibits the use of buried or excavated cisterns. Waste from such cisterns would not only be non-apparent, but would be difficult of detection.

No. 17 forbids the use of wooden receptacles not having proper metallic linings, *e.g.* water-butts. This regulation has the double object of preventing waste and avoiding pollution.

No. 18 requires the use of sound and suitable draw-taps, which must be of the "screw-down" kind.

Draw taps are divided into two classes, "plug-taps" and "screw-down" taps. In plug-taps the spindle or plug simply revolves in the tap, without rising or falling; a horizontal hole through the plug being made to connect or disconnect the inlet and the outlet to the tap by revolving the plug. This form of tap has no washer. The objections to the plug-tap are twofold, the most important being the sudden check which is given to the momentum of the body of water behind it, when the tap is closed. An illustration of the enormous strain upon a system caused by the use of plug-taps is attributed to Mr. A. R. Binnie (*Builder*, July 7, 1894, p. 3). In this experiment the pressure on a ¾-inch pipe, 114 feet long, and branching off a supply main, and furnished at the end with a plug-cock measuring 0·152 of an inch, was at the branch 120 lbs., and at the open cock itself 20 lbs. On the cock being shut quickly, those pressures were for the moment found increased to 220 lbs. at the branch, and 550 lbs. at the cock. The second objection to the plug-tap is its liability to leak, on account of the rapid wear, necessitating the plug being ground, which can only be done by a mechanic.

"Screw-down" taps (Fig. 52, Chap. XX.) have a screwed spindle, at the lower end of which is a washer, which, when forced against its seat by turning the spindle, closes the inlet to the tap. When a tap of this kind commences leaking, the old washer should be removed and a new one put in its place. This process is very simple and inexpensive, and can be performed by any intelligent person. The washers consist of leather, except for hot water, when vegetable fibre should be used. A tap invented by Lord Kelvin has been introduced within the last few years, in which the washer is constructed of gun-metal, and revolves upon its seat. In its earlier form it was not found to work satisfactorily under high pressure, but it has recently been much improved.

No. 19 refers to taps for " stand-pipes," and requires that they shall be of the " waste-preventer " kind, and be protected from injury by frost, theft, or mischief. With low-pressures automatically closing taps may be employed, but under high pressures they are rarely free from leakage. They are, in nearly all cases, open to the same objections as plug-taps.

No. 20 requires boilers, urinals, and water-closets to be served only through a cistern or service-box, and forbids the use of stool-cocks, or any direct communication between the company's pipes and such apparatus.

No. 21 requires the cistern supplying a " water-closet " to be fitted with a " waste-preventing " apparatus, capable of discharging not more than 2 gallons at each flush.

The desirability of altering this regulation being brought before the Local Government Board by several of the metropolitan sanitary authorities, the matter was referred to the London County Council, who again applied to the

M

Sanitary Institute for an opinion upon the subject. After much consideration and experiment, the Sanitary Institute reported as follows:—

"That Clause 21 of the Regulations under the Metropolis Water Act, 1871, should be altered to read, ' so constructed as to discharge not less than 3 nor more than $3\frac{1}{2}$ gallons of water at each flush.' "

The Local Government Board has, however, pointed out that such a recommendation was not within the purport of the Regulations referred to, which have for their object the prevention of waste, misuse, or contamination of water. The London County Council have agreed to make application to the water companies to amend the Regulations in the direction desired.

An unnecessary use of water is caused when water-closets adapted for use as urinals are fitted with waste-preventing cisterns discharging 2 gallons of water at each flush. For the latter use a much smaller flush would suffice, but there is no means of reducing the flush on such occasions.

No. 22 is similar to the last, and refers to urinals, but fixes the maximum flush at 2 gallons instead of 1 gallon.

No. 23 requires the "down-pipe" of a closet to have an internal diameter of not less than $1\frac{1}{4}$ inch, and if of lead to weigh not less than 9 lbs. to every lineal yard. The object of this, and Regulation 29, from the company's point of view, is not obvious.

No. 24 forbids any communication between the pipes supplying the company's water and any part of a water-closet, or any apparatus connected therewith, except the service cistern.

No. 25 prohibits the existence of any overflow pipe to a bath, unless it be constructed as a "warning-pipe." The remarks made in connection with Regulations 14 and 15 apply to this regulation, but to a modified extent.

No. 26 requires that the inlet shall be distinct from, and unconnected with, the outlet of a bath ; that the inlet shall be above the highest water-level of the bath ; and that the outlet shall be provided with a perfectly water-tight plug, valve, or cock. These requirements aim at the abolition of baths having a combined inlet, outlet, waste, and overflow (the last being, however, prohibited by Regulation 25) at the bottom. These arrangements are very liable to get out of order, and water might easily escape directly into the drain without ever being noticed.

No. 27 requires that no alteration shall be made in any fittings in connection with the supply of water without two days' previous notice in writing to the company. Compare Waterworks Clauses Act, 1863, sec. 19.

No. 28 is for the protection of the consumer, where the communication-pipe is laid by the undertakers, and sets forth that no cock, ferrule, joint, union, valve, or other fitting, in the course of any " communication-pipe," shall have a less water-way than that of the communication-pipe.

No. 29 requires that all " warning-pipes " and other lead pipes of which the ends are open, so that such pipes cannot remain charged with water, may be of the following minimum weights :—

Internal diameter in inches.	Weight per yard in lbs.
$\frac{1}{2}$	3
$\frac{3}{4}$	5
1	7

No. 30 defines " communication-pipe " as being the pipe which extends from the district-pipe or other supply-pipe of the company up to the " stop-valve " prescribed in Regulation 12.

No. 31 imposes a penalty of **£5 on any person contravening** these Regulations.

No. 32 empowers an authorized officer to act for the company.

No. 33 states that all existing fittings approved by the company shall be deemed to be prescribed fittings under the Metropolis Water Act, 1871.

CHAPTER XXII

STORAGE OF RAIN-WATER.

As the rainfall is stored on a large scale by impounding reservoirs for the purposes of providing a supply of water for domestic and other purposes to towns or a series of villages, so also it may be collected and stored on a small scale for the use of isolated dwellings. In the latter case, roofs or prepared areas of ground take the place of the catchment area; the impounding reservoir is replaced by metal, masonry, brickwork, or concrete tanks, which also constitute the service-tanks; and filter-beds of adequate dimensions have also to be provided. Storm overflow must not be omitted, but in this case with the sole object of preserving the purity of the water in the tank. ;The principles involved in the necessary calculations are likewise similar, the supply depending upon the available rainfall and the area of the collecting surfaces.

Rain-water from the roofs of houses is usually collected by means of gutters, with the immediate object of preventing it from flowing down the walls and rendering them damp. It is also usually stored in butts or tanks for washing purposes, when the principal supply is of a hard nature, and for other purposes when the latter is limited in quantity. It is rarely, however, employed for dietetic purposes, except as a last resource, when no other supply is forthcoming. In fact, the prejudice against it is so great that people frequently prefer to it the water of wells or springs which they know to be polluted. Another reason, perhaps, is the general ignorance of the capabilities of a comparatively small collecting surface in affording a regulated daily supply through the use of storage capacity properly proportional to the particular area of surface and the available rainfall.

As stated in Chapter IX., the amount of rainfall varies considerably in different parts of the country, from 40 to 70 inches per annum on the western coast, from 30 to 40 inches on the southern coast, and from 20 to 30 inches on the eastern coast—the average rainfall over the whole of Great Britain being about 33 inches. Further, it was stated that the rainfall at the same place varied considerably in different years—from 45 per cent. above the average to 33 per cent. below the average. There is, however, another most important variation in the rainfall, when corresponding months in different years are compared, which plays a most important part in the calculation of the necessary amount of storage-capacity for roof or similarly collected supplies. In this case there is no percolation, and the rainfall passes directly into the storage tanks, the flow not being subject to regulation by the catchment area.

An examination of the table given on p. 166, taken from "Sanitary Engineering," by Baldwin Latham, M.I.C.E., F.G.S., will show that, although the monthly averages are fairly constant when the observations extend over several years, the individual falls vary from 204 per cent. above the average, as in December, 1876, to 90 per cent. below the average, as in September, 1865.

The theoretical storage capacity should be such that the tank should contain a sufficient quantity of water at the commencement of a drought so as to afford the calculated daily supply until it is over; also to be able to store the surplus of the rainfall above the consumption during wet weather. Or, in other words, that there shall always be a sufficient supply of water in the tank—provided that only the calculated amount is abstracted from it daily—and that the tank shall never overflow. The lowest annual rainfall must be taken as the basis upon which the daily supply is to be calculated. In the above table the lowest annual rain-fall was 20·26 inches, which, if properly stored (excluding

RAINFALL AT CROYDON, COMPILED FROM OBSERVATIONS MADE BY MR. GEO. CORDEN, CROYDON.

Rain-guage, 154·6 O.D.　"Sanitary Engineering," 2nd edit., by Baldwin Latham, M.I.C.E., F.G.S., pp. 54 and 55.

Month.	Monthly average.	1861	1862	1863	1864	1865	1866	1867	1868	1869	1870	1871	1872	1873	1874	1875	1876	1877
January ...	2·95	·49	2·10	2·87	1·34	3·45	4·51	3·12	4·31	2·91	1·43	2·87	5·54	4·08	1·15	3·49	1·04	5·53
February ...	1·70	2·37	·48	·61	1·32	1·82	5·11	1·24	1·22	2·46	1·72	1·19	1·07	1·69	1·83	1·04	2·05	1·70
March ...	1·82	2·72	3·26	·72	2·95	1·05	2·08	2·58	1·08	1·52	1·84	1·50	2·13	1·55	·55	·65	2·57	2·16
April ...	1·54	·61	2·27	·61	·64	·27	2·04	1·75	1·92	1·16	·42	3·23	1·23	·63	1·87	1·74	1·67	4·10
May ...	1·82	·93	3·45	1·50	2·69	3·40	2·17	1·52	·76	3·20	·86	1·00	3·05	1·21	·86	·95	·93	2·39
June ...	1·93	1·90	3·06	3·59	·80	1·99	3·05	1·85	·35	1·34	·20	2·68	1·96	3·41	2·54	2·26	·99	·87
July ...	2·25	2·78	2·09	1·16	·99	3·33	1·48	4·21	2·23	·66	2·24	2·89	3·29	2·27	1·18	4·50	·34	2·58
August ...	2·15	·54	3·36	2·10	1·19	3·69	3·16	2·08	2·98	1·24	1·86	·81	2·24	2·53	2·41	·99	2·75	2·70
September...	2·69	2·16	1·96	5·16	3·09	·27	4·07	2·57	1·71	3·32	2·44	5·00	1·63	2·30	2·49	2·90	3·07	1·65
October ...	2·93	1·03	4·92	2·22	1·26	7·20	1·33	1·89	2·37	1·66	4·24	1·17	5·24	3·14	4·83	4·07	1·31	1·97
November...	2·53	5·41	1·37	2·08	3·46	3·14	1·38	·62	1·15	2·53	1·89	·55	3·50	2·54	2·59	3·15	2·57	4·94
December...	2·44	1·29	2·02	2·88	·53	1·80	2·00	1·61	4·50	3·38	3·03	1·65	4·39	·36	1·80	1·22	7·43	1·61
Total annual average.	26·75	22·23	30·34	25·57	20·26	31·41	32·38	25·04	24·58	25·38	22·17	24·54	35·27	25·71	24·10	26·87	26·82	32·1x

loss by evaporation) would supply a monthly consumption equal to 1·69 inches. The storage capacity necessary to equalize the rainfall in order to furnish this supply may be calculated by the method given in Chap. XIII., either graphically or arithmetically.

The calculation of storage capacity based upon the rainfall for 1864, given in the table above, is as follows:—

Month.	Rainfall.	Consump-tion.	Algebraical sum.	Accelerated sum.	A
January ...	1·34	1·69	− ·35	− ·35	1·25
February ...	1·32	1·69	− ·37	− ·72	·88
March ...	2·95	1·69	+ 1·26	+ ·54	2·14
April ...	·64	1·69	− 1·05	− ·51	1·09
May ...	2·69	1·69	+ 1·00	+ ·49	2·09
June ...	·80	1·69	− ·89	− ·40	1·20
July ...	·99	1·69	− ·70	− 1·10	·50
August ...	1·19	1·69	− ·50	{ minimum } − 1·60	0·00
September	3·09	1·69	+ 1·40	− ·20	1·40
October ...	1·26	1·69	− ·43	− ·63	·97
November...	3·46	1·69	+ 1·77	{ maximum } + 1·14	2·74
December...	·53	1·69	− 1·16	+ ·02	1·58

To give an equalized monthly supply equivalent to 1·69 inches would, therefore, require a minimum storage capacity for 1·60 + 1·14 = 2·74 inches, provided that the tank contained at the commencement of the year a quantity of water equivalent to 1·60 inches.

The column marked A in the above table shows the depth of rainfall corresponding to the quantity of water contained in the tank at the end of each month, under these conditions. This is assuming that the rainfall is uniform throughout each month, but as this is not usually the case, and the bulk of it may fall during the first day or two, it is necessary to increase the storage by one month's consumption = 1·69 inches, making the total storage capacity equal to 4·43 inches, or about 22 per cent. of the total annual rainfall. Mr. Thomas Box ("Practical Hydraulics") gives one-fourth of the total annual rainfall as the minimum storage capacity, and this may be taken as a safe rule.

There are several rules for calculating the supply of water obtainable from a given collecting surface and annual rainfall, three of which are given here :—

1. Multiply the area in square feet by half the annual rainfall in inches = the quantity of water in gallons approximately (Parker's " Practical Hygiene ").

2. Multiply the area covered by the roof in square feet by the average annual rainfall, also in feet, and divide the result by 100 = the average supply in gallons per diem, for a very dry year (J. Wallace Peggs, " Transactions," Sanitary Institute, vol. xiv.).

3. Multiply the lowest recorded rainfall in feet by the collecting area in square feet by ·015 = the average number of gallons per day, including a small allowance for loss in the collection ("Water, its Composition, etc.," by Joseph Parry, C.E.).

The following table, by W. Sowerby Wilson, F.R.M.S., shows the daily yield of water from a roof with varying rainfall: area of house, 10 feet by 20 feet, or 200 square feet :—

Mean rainfall.	Loss from evaporation.	Requisite capacity of tank.	Mean daily yield of water.	Mean daily yield of water in wettest year.	Mean daily yield of water in dryest year.
Inches.	Per cent.	Cubic feet.	Gallons.	Gallons.	Gallons.
20	25	100	4·3	6·7	3·2
25	20	135	5·7	7·5	3·9
30	20	145	6·8	9·4	4·5
35	20	155	7·9	11·0	5·0
40	15	165	9·7	13·1	7·2
45	15	170	10·9	14·2	8·6

For any other size of roof or amount of rainfall, the numbers will be proportional.

The area of roof-surface for collecting water must be measured on the flat, and not on the slope, and is the same as the area of the ground covered by the roof (if the ground be level). The average collecting area of house-roofs is about sixty square feet per individual.

Roofs of houses in towns, or near factories, or in any situation where the air is liable to pollution, are not suitable for collecting water for dietetic purposes, and roofs covered with thatch, felt, lead or zinc—the two last, on account of the solvent property of rain-water,—must not be used for the purpose. Lead pipes in connection with the storage-tank must also be avoided. In order to obtain a satisfactory supply of water for domestic purposes from roofs or similar areas, the greatest constant care is necessary to keep the roofs and gutters clean, and to remove all bird-excrement, etc.

The first portion of the rainfall collected from roofs, especially after a long period of dry weather, is usually of an impure nature, on account of the soot and other matters which have had time to accumulate. In order to prevent this portion from entering the storage tank, mechanical appliances, known as "rain-water separators," have been devised. These consist of vessels into which the rainfall is directed before passing into the storage tank, and which are so constructed as to collect the first portion of the flow, and when full to cant over, emptying their contents to waste, and at the same time placing the flow of water from the roof in direct communication with the storage tank. When the flow has ceased they return automatically to their original position.

The rain-water may be stored in overground or underground cisterns, or reservoirs. The former have the advantage of enabling the water to be drawn direct by means of a tap, but are generally more expensive, and the water is subject to changes of temperature. The underground reservoir necessitates pumping, and is more liable

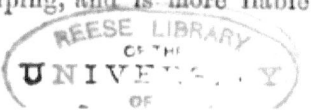

to permit of pollution. The water should be strained

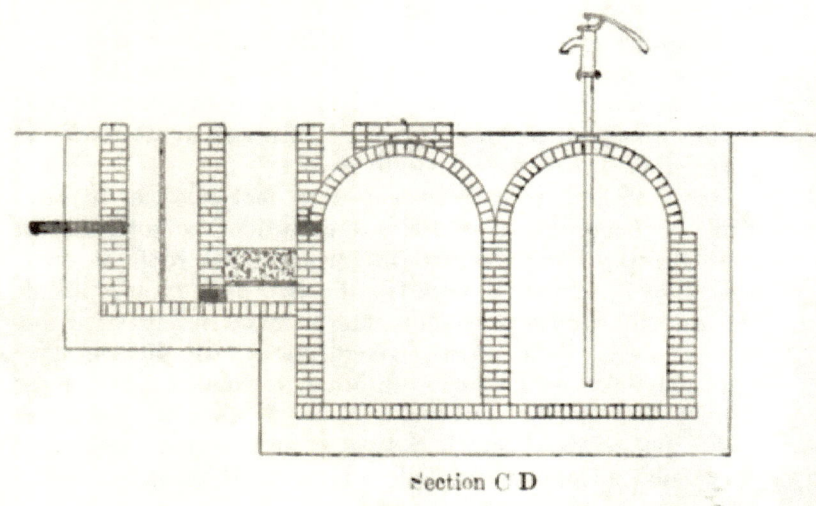

Section C D

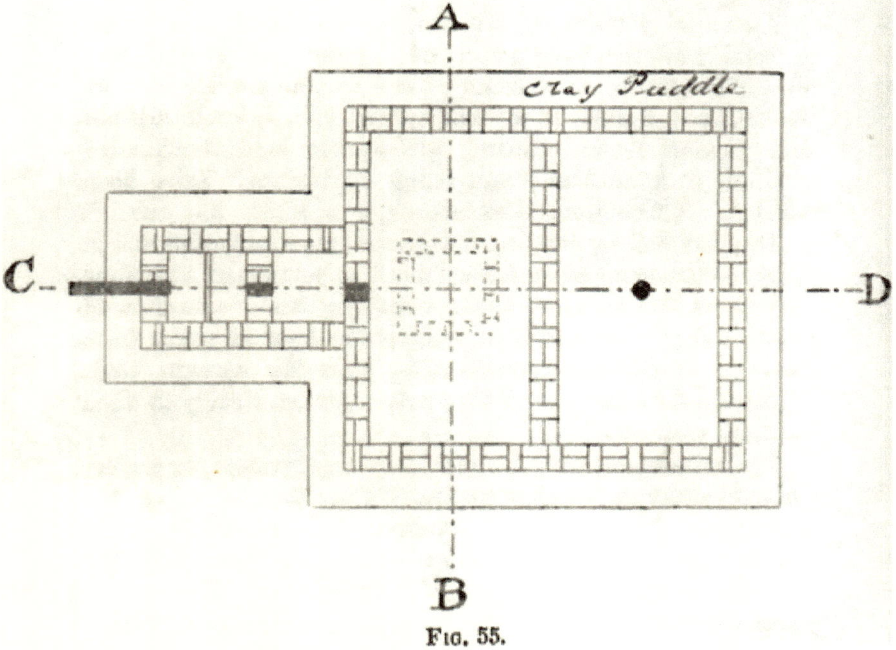

Fig. 55.

through a copper strainer to keep back leaves, etc., and should be filtered before admission into the storage tank. The filtering material may be sand, polarite, or other suitable substance. The tank should be efficiently ventilated; have a sufficient covering of earth so that the water may be kept at an equable temperature; and should be, if possible, provided with a wash-out valve for cleansing purposes. Fig. 55, is an illustration of a brick underground tank, with strainer and filter-bed, to contain 2500 gallons, designed for a small isolation hospital.

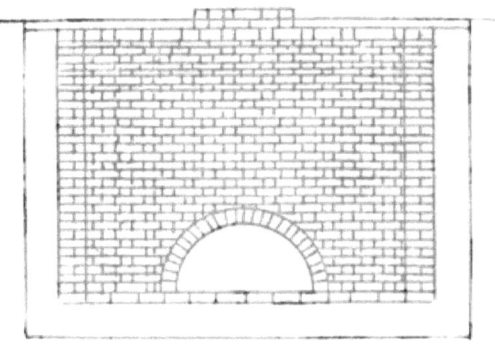

Section A B

FIG. 56.

Where the roof surface is insufficient, a small area of ground may be fenced off, and either underdrained by means of agricultural tiles, or the surface may be covered with tiles or concrete, and the rain-water falling upon it taken by means of pipes to a storage-tank.

CHAPTER XXIII.

SPECIFICATIONS AND ESTIMATES.

THE preparations of the drawings, specifications, and general conditions of an intended work should receive the greatest care and attention, the descriptions being in detail and clearly expressed, otherwise the contractor will construe them in one way and the engineer in another. It is therefore in the interests of both parties to the contract that sufficient time and care should be taken in drawing up the clauses and in giving as much information as possible to the contractor. This will avoid disputes during the progress of the work, which usually result in claims for extra payments over and above the contract amount. The bill of quantities showing the amount of work to be performed under each item of the specification is either prepared by the engineer or by a quantity surveyor, for which it is customary to allow from $\frac{1}{2}$ to 1 per cent. on the amount of the accepted tender. This is charged through the bill of quantities, and repaid or transferred to the engineer or quantity surveyor by the contractor. It is not a desirable arrangement for the engineer to have any monetary transactions with the contractor, and it would undoubtedly be an advantage where the quantities are taken out by the engineer for this payment to be made direct to him by the person or persons for whom the work is being performed.

The following specification for cast-iron pipes will act as a guide to the preparation of contracts for such works, at the same time giving an insight into the style and method usually adopted in the preparation of specifications generally. The general conditions in this case are embodied in the specification, and are not kept distinct, as is usual in large contracts for engineering works.

Rural Sanitary District of ——.

Cast-iron Pipes, Irregulars, and Special Castings.

Specification to be observed by the contractor for the supply of ordinary iron pipes, irregular iron pipes, and special castings, to the Waterworks Department of the Local Authority of the Rural Sanitary District of ——.

1. The cast-iron pipes, from 3 inches to 12 inches in diameter, are to be in 9 feet lengths, and the 2-inch pipes are to be in 6 feet lengths, in each case exclusive of the socket. The whole or any portion of the above sizes to be spigot and socket, or half-turned and bored joints as may be directed to be supplied, and to be of the following weights, except when otherwise ordered.

Diameter in inches.	Length, exclusive of socket.	Weight, including socket.			Diameter in inches.	Length, exclusive of socket.	Weight, including socket.		
ins.	ft.	cwts.	qrs.	lbs.	ins.	ft.	cwts.	qrs.	lbs.
2	6	0	1	19	8	9	3	2	10
3	9	1	0	7	9	9	4	2	6
4	9	1	1	15	10	9	5	0	3
5	9	2	0	2	11	9	5	2	1
6	9	2	1	14	12	9	6	2	22
7	9	3	0	19	—	—	—	—	—

2. Any pipes which deviate more than 3 per cent. from the stipulated weights will be rejected. The whole of the pipes are to be manufactured and afterwards tested by hydrostatic pressure, by and at the expense of the contractor.

The test pressure to be equal to a column of water 600 feet high, and such pressure shall be maintained in each pipe at least two minutes, previous to which the connection between the pump and the testing-machine is to be cut off. Each pipe, while under the pressure, shall be rapped from end to end with a hand-hammer 4 lbs. in weight, so as to discover any sandy, porous, or blown places. The pipes will be again proved by and at the expense of the authority, after they have been delivered at the place required. Any pipe which shall be found to be imperfectly coated, or in which any imperfections shall appear, or wherein any sand or air-holes shall appear to have been plugged up, or which shall not agree with the terms of the specification, will be rejected.

3. The irregular pipes and special castings are to include all branches, elbows, thimbles, clips, cant socket-pieces, hydrant, valve, meter, and stop-tap covers fitted as per pattern, also all flange and other special castings, samples of which may be seen on application at the waterworks offices.

4. The straight pipes are to be cast in dry sand moulds vertically, with the sockets downwards, and the curved pipes in loam or dry sand in close boxes; the castings are to be made without the use of core-nails, chaplets, or thickness pieces, or any substitute for the same, and the contractor is to provide turned iron patterns, boxes, core-bars, and barrels for making all straight pipes, the flanges, spindles, sole plates, and cores of which are to be accurately turned, faced, bored, truly fitted and joined. The sand must be sufficiently fine and fresh to produce a smooth and perfect surface, and all the moulds and cores are to be properly black-washed and carefully dried. Great care is to be taken in preparing and drying the cores in order to ensure a smooth surface to the pipes internally.

5. All pipes of 5 inches diameter and under will be allowed to be cast at an angle of not less than 45 degrees from the horizontal. The pipes of each size respectively are to be of uniform bore and thickness of metal through-

out their respective length, and without any belts. The castings are to be free from scoriæ, sand-holes, air-bubbles, cold-shuts, laps, washers, and all other imperfections; and the pipes are to be truly cylindrical in the bore, straight in the axis, smooth within and without, and internally of the full specified diameters, and they shall have their inner and outer surfaces as nearly as possible concentric. An increased length of at least 6 inches is to be cast on the spigot end of each pipe, such increased length being afterwards cut off in a lathe to the specified size. All pipes are to be perfectly dressed and cleansed, so that no lumps or rough places are left in the barrels or sockets. The contractor will be charged with and must pay and defray any and all losses, charges, and expenses to or for which the Local Authority or their Committee may be put, or be liable by reason of any neglect with respect to the forms and sizes of the sockets and spigots. And for the better prevention of chills, uneven shrinkage, and cracks consequent thereon, the contractor must undertake that the pipes shall not be removed too hastily from the moulds, or be laid while hot upon cold or damp earth, or be exposed while in a heated state to wet or inclement weather.

6. Special precautions are to be taken in obtaining and maintaining the proper standard of quality of the mixture of iron, and in the melting of it. All the metal used shall be made from mine pig, without any admixture or proportion of cinder iron or other iron of inferior quality; and the whole to be of the best tough, close-grained grey iron, to be remelted in a cupola or air-furnace. Whenever any castings are made, as many test-bars are also to be cast from the same metal as may be required by the Local Authority or their engineer. The test-bars are to be 2 inches by 1 inch by 3 feet 6 inches long, placed horizontally upon the narrow width, supported on bearings 3 feet apart, and to be capable of sustaining without breaking a weight of not less than 30 cwt., gradually applied in the centre of the bar, and producing a deflection of not less than from ·3 to ·4 of an inch. The contractor is to provide in his works, under cover and

protection, suitable and approved machines for testing the test-bars to the full satisfaction of the engineer before he will be allowed to commence work under this contract.

7. All pipes which are ordered to have half-turned and bored joints are to be cast with such increased thicknesses of metal at the spigot and socket ends as will allow them to be so turned and bored and otherwise finished as that the spigot end of any pipe shall enter to the bottom and fit perfectly in the socket of any other pipe, and that when driven together, the whole of the turned and bored portions are in complete contact with each other, and form a perfectly water-tight joint, without the aid of any cement, lead, paint, or other substance. The turning and boring is to be executed after the pipes are coated as hereinafter specified. The turned portion of the spigot is to be ¾ of an inch in length, with a taper of 1 in 32, the socket having a similar taper to suit it. The spigot and socket joints, when so ordered, are to have a recess cast around the inside of the socket, and a bead on the spigot according to the directions of the Local Authority or their engineer. The sockets of all pipes under 4 inches in diameter are to have an allowance of ¼ of an inch, and above 4 inches in diameter an allowance of ⅜ of an inch for the lead-joint.

8. Each pipe is to have a consecutive number, year (initials of authority), and maker's name cast on the socket, in the same order, as specified on each order for goods, and any pipe found to be defective under the tests applied shall not have its number replaced, such number on the defective pipe being disfigured by a chisel-cut.

9. All pipes, irregulars, and special castings are to be heated in a stove to a temperature of 300° Fahr., and then dipped into a solution known as Dr. Angus Smith's, which must be of a similar temperature to the pipes.

10. All irregular pipes and special castings are to be made according to particulars to be furnished hereafter by the engineer; but all patterns and moulds that may be required in connection with the above are to be provided at the cost of the contractor.

11. The engineer or any other person the Local Authority may appoint will be empowered to reject any pipe or other casting he may consider defective, unsound, badly varnished, of inferior quality, or not in accordance with the order, and the same is to be removed by the contractor after notice has been given to him, at his own expense; and if not removed within three days after a written notice has been served upon him, the whole or any portion of such defective material may be removed by the Authority at the contractor's expense.

12. Orders signed by the engineer or other authorized person for the Local Authority will be given from time to time for such descriptions of pipes and other articles as may be required under the contract, to be delivered within the following periods of time:—

Straight pipes, 2 to 8 inches in diameter, inclusive: 100 within 14 days, and 100 per week afterwards.

Straight pipes, 9 to 12 inches in diameter, inclusive: 50 within 14 days, and 50 per week afterwards.

Branch pipes, and other irregular or special castings, 2 to 8 inches in diameter, inclusive: 10 within 14 days.

Branch pipes, and other irregular or special castings, 9 to 12 inches in diameter, inclusive: 10 within 14 days.

13. Should the contractor fail to deliver the pipes, irregulars, or special castings as required, the Local Authority shall have power to determine or cancel the whole or any portion of the contract, or to order elsewhere any goods not supplied, and the difference in cost between the goods so supplied and the contract price may be charged to the contractor, or deducted from any amount due or to become due to him under the contract.

14. The pipes and all other castings are to be delivered to the ———— station of the ————Railway Company, when delivered by rail, or at the waterworks yard when delivered by road or canal, free of charge to the Local Authority, and the said Authority reserve to themselves the right to weigh all pipes and castings on their own weighing machines, and to pay for same according to such weights.

N

15. Should any dispute or difference of opinion arise as to the meaning or intention of this specification or of the contract, the interpretation of the same by the clerk to the Local Authority or other person to be agreed upon before entering on this contract, shall be binding and conclusive on both parties in the matters to which such interpretations shall refer.

16. The Local Authority do not bind themselves to accept the lowest or any tender, and reserve to themselves the right to divide and accept part or parts only of a tender.

17. The contractor must enter into a bond for the due fulfilment of the contract under the penalty of £100, such bond and agreement containing the written terms and stipulations to be prepared by the clerk to the Authority, and sealed by the said Authority.

18. Payments will be made monthly after the next succeeding meeting of the Authority, if accounts are sent to the engineer on or before the ———— day of any month.

19. Tenders are to be made on the annexed form, four prices per ton to be given, viz. :

    (1) For ordinary cast-iron pipes—

        a. Spigot and socket.

        b. half-turned and bored.

    (2) For irregular castings.

    (3) For all other special castings.

Form of Tender.

Address ——,

Date ——.

To the —— Rural Sanitary Authority.

Gentlemen,

    I beg to tender for the following *pipes, irregulars, and other castings,* as per specification, for twelve months from the above date.

£ *s. d.*

Ordinary cast-iron pipes (spigot and socket) per
ton   ...   .

Ordinary cast-iron pipes (half-turned and bored)
per ton   ...   .

Irregulars as under—

Taper or reducing pipes   ...   ...
Thimble or other branches   ...   ...
Saddle branches   ...   .
Bends, elbows, and cant sockets   ...
Thimbles or collars   ...   .
Clips ...   .
Cap ends   ...   .
Duckfoot bends   ...   .
Flange pieces (various)   ...   ...
Blank flanges ...   .

} per ton

Special Castings as under—

Hydrant covers, as per sample, fitted
Meter   „   „   „   „
Stop-tap   „   „   „   „
Valve   „   „   „   „
Air-valve   „   „   „   „
Meter caps   „   „   „   „
False spindles of various lengths   ...

} per ton

Signed ——,

Address ——.

In preparing estimates of work to be performed, careful consideration must be given to the district in which the work is to be executed; and the correct estimation of the prices requires considerable experience of practical work. The following statement is the actual cost of laying a 3-inch main along a main road, giving the details and prices of each item. All estimates should be taken out in a similar manner.

DISTANCE 117 YARDS—3-INCH MAIN.

No.		Tons.	cwts.	qrs.	lbs.		£	s.	d.
39	3-in. pipes ...	2	1	—	—	at £5 per ton ...	10	5	0
1	5 by 3-in. branch	—	1	1	14	at 10s. 1½d. per cwt.		13	10
2	3 by 3-in. branch	—	1	—	—	„ „		10	1
1	3-in. bend ...	—	—	2	7	„ „		5	9
1	3-in. valve ...	—	—	—	—	... ...	1	16	1
1	Valve cover ...	—	—	2	7	at 16s. 1¼d. per cwt.		9	2
—	Lead ...	—	—	1	—	at 12s. per cwt. ...		3	0
—	Yarn ...	—	—	—	2	at 2d. per lb. ...		0	4
—	Coal ...	—	1	2	—	at 9d. per cwt. ...		1	2
—	Cement ...	—	1	—	—	at 2s. per cwt. ...		2	0
—	Hauling ...	2	—	—	—	at 1s. 3d. per ton ...		2	6
	Wages paid for labour						7	0	0
							£21	8	11

CHAPTER XXIV

WELLS.

WELL-SINKING on a large scale for rural water-supply is of rare occurrence, and the details, which necessitate an extensive knowledge of practical geology, are so numerous and complicated that the subject cannot be dealt with satisfactorily within the limits of this book. The principles and methods have been comprehensively dealt with in several standard treatises upon the subject, and to these the student must be referred. As, however, wells may be said (and often unfortunately so) to form the most frequent source of supply for villages, groups of houses, and isolated dwellings in country districts, some details upon this branch of the subject will be given here.

Wells may be classed under three heads—

1. Shallow or surface wells (Fig. 56).
2. Deep wells (Fig. 57).
3. Artesian wells (Fig. 58).

In Chapter II., whilst referring to the disposal of the rainfall, it was stated that "a portion of the rain sinks into the ground and forms the underground reservoirs in which wells are sunk, issuing again at the lowest lip as springs." When a porous stratum, such as sandstone or chalk, which has the power of retaining water in its pores or fissures, is superposed upon an impervious stratum, such as clay, the porous stratum will be saturated, and the water held up as in a basin, to a plane inclining towards the lowest lip, which is generally the outcrop of the impervious stratum.

If the porous stratum is adjacent to the surface of the ground, the plane of saturation is generally at no great depth; and if a well is sunk to a point below this plane, water will collect in it and stand at the level of the plane. This constitutes a shallow well. Under these circumstances the level of the plane of saturation is very variable, being rapidly affected by the rainfall. As the rainfall in its

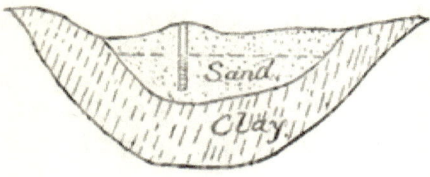

FIG. 57.

passage into and through the ground, on account of its highly solvent nature, takes up and carries with it any impurities, more particularly of an organic nature, which it meets with on its way; and as the distance through which it has to pass before flowing into the well is usually very short, shallow wells are dangerous sources of water supply for domestic purposes. It is only in such cases as where

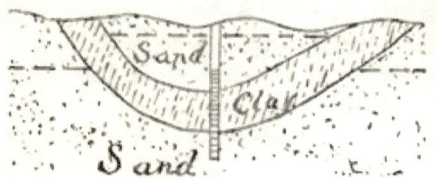

FIG. 58.

the well is in an isolated position, in a rural district, and sufficiently removed from any possible source of pollution, that its use in this connection should be permitted. Wells of this description must be properly walled in or steined with stone, brickwork, or concrete, and in the two first cases the joints should be made thoroughly watertight with hydraulic lime-mortar or cement. The top of the wall should be protected by a raised kerb, where a bucket is used for drawing the water, and fitted with a proper cover, or where the water is drawn by means of a pump, the top of the well should be domed or flagged over. The bucket system has an advantage over the pump in affording greater facilities for cleaning out the well, which should be done once a year. On the other hand, the pump-well, through being permanently covered over, is less liable to pollution from the surface. As the water in shallow wells is usually of a soft nature, the suction pipe of the pump should not be composed of lead.

Deep wells are those which are sunk through an impervious stratum to a porous or water-bearing stratum

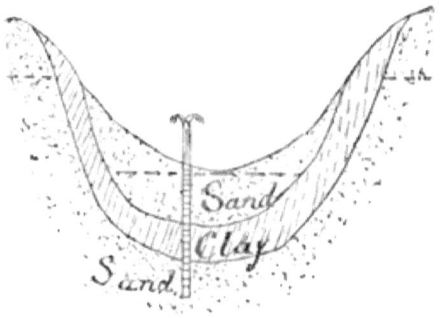

FIG. 59.

beneath it, the water being held up in the latter by an impervious stratum underneath it again. The terms "deep" and "shallow," in connection with wells, do not, therefore, refer to the actual depth of the well, and a shallow well may, in fact, be deeper than a deep well. Deep wells, if properly constructed, constitute excellent sources for domestic supply. The rainfall which feeds them is collected upon the exposed surfaces of the water-bearing stratum, which are usually situated at a distance from the site of the well, and becomes purified in its passage through the ground. On account of its prolonged contact with strata at some depth below the surface, deep well-water usually contains a considerable

amount of mineral matter in solution which it has taken up
during its passage; this gives it a hard character. Deep
wells have also the advantage of being slowly affected by
the rainfall, and the level of the water in them is fairly
constant. It is of paramount importance that any percolation from the beds above the impervious bed through
which the well is sunk should be effectually prevented.
This is done in a manner similar to that described for
shallow wells. The precautions at the surface are the same
in both cases. The level at which the water will stand in
a " deep well " depends upon the elevations of the collecting
ground and the line of overflow, the principles upon which
it depends being the same as already described in reference
to the virtual slope or hydraulic mean gradient of water
flowing in pipes.

There is a continuous flow of water in saturated strata
from the collecting area towards the outlet, which is usually
the bed of a river, or the shore of a lake, or the sea. The
surface-level of this moving body of water, which may be
called its virtual slope, depends upon the resistance of the
materials which compose the strata through which it flows,
the presence of faults or dislocations, and the physical
features of the land ; technically, the first of these includes
the other two.

Should the point selected for sinking a deep well be
situated beneath the virtual slope of the water in the saturated beds, then, when these beds are reached, the water
will rise to the top of the well and (were it not for the
resistance of the air) above it to the virtual slope at that
point. This would constitute an artesian well. The name
is derived from Artois, a province of France, where this
form of well was first brought into general use. It will be
evident to the student that the artesian well is only a special
condition of the deep well. As in wells of this description
the water rises of its own accord, either so as to overflow,
or to within a certain distance from the surface, it is only
necessary to dig the well to a sufficient depth to allow of
the pump being fixed within 30 feet of the lowest level to

which the water rises, and to afford sufficient storage
capacity. The remaining portion may consist of a small
perforation bored down to the required depth, which is
lined with an iron tube, or occasionally left unlined when
it passes entirely through rock.

The Abyssinian or tube-well (Fig. 60) is economical and
satisfactory where the ground is suitable, and where the
water stands, or by deeper sinking may be made to rise
within 30 feet of the surface of the ground. This well
consists of iron tubes from 1¼ inch
diameter, in sections, which are
driven into the ground, the bottom
section, which is perforated, having
a steel point to enable it to pene-
trate. As the tube is forced down
into the ground, a fresh section is
screwed on to the upper end of the
last tube until the desired depth is
reached. A pump is then fixed to
the top of the composite tube, and
the work is complete. An advantage possessed by this form of well
is that it can usually be withdrawn
and driven again in a new situation.
Percolation of surface-water between the lining of the well, and
the ground through which it is
driven, is also prevented.

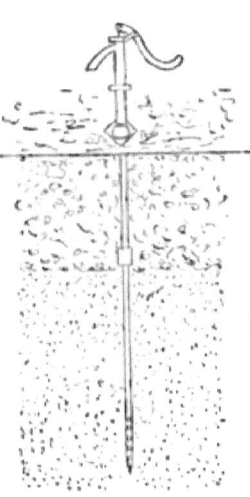

FIG. 60.

The following is an estimate for an Abyssinian tube-well
in gravelly ground, where driving-plant is provided and a
well-driver, sent by the contractors to superintend fixing,
labour being provided by the Authority.

ESTIMATE.

Depth, 30 feet		£	s.	d.
30 feet 1¼-inch Abyssinian tube, hire of plant, and well-driver	...	4	18	0
3-inch Abyssinian column-pump, with foundation	...	2	15	0
Well-driver's time, travelling, say one day	... ...	0	10	0
Ditto, ditto, railway fares	...	1	1	0
Carriage and Cartage of Plant and materials ...	...	1	0	0
Labourer, 1 day	...	0	2	6
		£10	6	6

For further information upon this subject, the student is referred to a pamphlet published by Messrs. Le Grand and Sutcliffe, Bunhill-row, London, E.C.

The cost of digging wells depends upon the nature of the rock or soil through which the excavation has to be made, and upon the precautions which have to be taken to prevent the sides from falling in, and in dealing with the surface water.

There are two principal methods of constructing the steining of a well in treacherous ground. The first is to excavate to a certain depth, and then to build the lining of the well to the surface of the ground upon a wooden ring or kerb. The excavation is then continued for a further depth, and the kerb, with the cylinder of brickwork, or other material, allowed to sink. The walling is then continued again to the surface, the excavation and walling being carried on alternately until the required depth is reached. Iron cylinders in sections are also used, the principle being the same. The second method consists in under-pinning and building beneath each section of the steining as constructed.

Half brick rings are usually sufficient, especially when laid in cement.

The courses should break joints, and the bricks should be radiating.

It is always advisable to have a puddle or concrete backing to the lining of a well, especially near the surface.

SPECIFICATION of a well and other works required to be constructed for Mr. ———, on land adjacent to the main road leading from ——— to ———, in the parish of ———, in the county of ———.

The works comprised in this contract are the excavation and lining of a well, and other works, as shown on the drawing (Fig. 61) attached to this specification, and hereinafter more particularly detailed and described.

The site of the well is shown on the aforesaid drawings, and the strata will probably consist of—

Alluvium ...	...	...	...	...	2 feet
Sand and gravel ...		...	...	...	4 ,,
Soft and hard marl		...	...	...	64 ,,
Sandstone ...	...	...	...	...	10 ,,

The well is to be excavated or sunk to a depth of 80 feet, and 4 feet 6 inches diameter in the clear inside the brickwork lining. The excavation is to be trimmed back of sufficient width for the lining and concrete backing, or 7 feet 6 inches in diameter where lining is inserted with concrete backing, and 6 feet in diameter where brickwork lining only is inserted; the remainder of the excavation, where lining is not found necessary, is to be made 4 feet 6 inches in diameter, and neatly trimmed to the circle.

The excavated material is to be removed by the contractor as part of this contract.

Blasting with explosives will be permitted, but no shots are to be fired within one foot of the sides of the excavation.

The brickwork lining is to be 9 inches in thickness, built solid in Portland cement mortar, and to consist of header and stretcher-courses, constructed in sections on the system known as "underpinning," but no section of the brickwork is to exceed one-fourth of the circumference, or 4 feet in vertical height as a maximum. Great care must be taken in measuring the height of the courses so as to avoid wide closing-joints. The brickwork is to be built in, as the

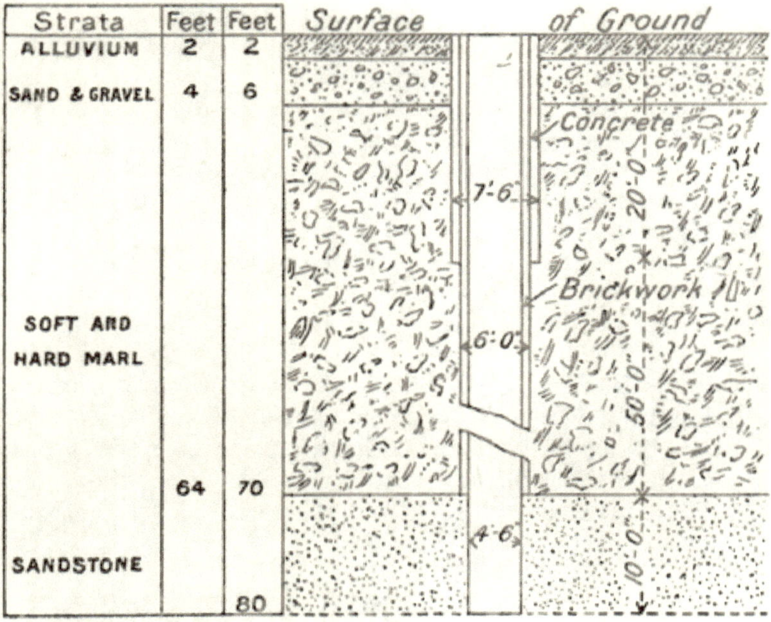

Fig. 61.

sinking proceeds, where the ground is not strong enough to stand alone. No vacant spaces are to be left behind the lining, but all such spaces must be filled in with fine cement. The lining is to be kept perfectly plumb, and worked from a radius-rod off the centre-line.

The ring of cement concrete backing to the brickwork is to be carried to a depth of 20 feet from the surface, and to be 9 inches in thickness, and well rammed into its position behind the brickwork.

The contractor is to provide, as part of this contract, all carriage, labour, materials, tools, pumping apparatus, or other means of lifting the water during the execution of works, and any other apparatus necessary for the due and proper execution of this contract.

A hand windlass of elm-wood, with substantial standards and frame, is to be provided and fixed at the top of the well, after the sinking and other operations have been completed. A best Manilla rope, 2½ inches in circumference, with swivel attachment, and a strong elm or oak bucket, holding not less than 3 gallons, are also to be provided as part of this contract.

The works are to be completed to the satisfaction of Mr. ——, within two months after the date of the signing of this contract.

Payments will be made weekly, at the rate of 80 per cent. of the total work executed.

MATERIALS.

The bricks are to be of approved manufacture.

The cement-mortar is to be composed of one part of Portland cement, and three parts clean furnace ashes or sharp sand.

The cement concrete is to be composed of one part of Portland cement, one part of clean, sharp sand, and three parts of small broken stone.

The timber is to be of the best-seasoned elm, free from all imperfections.

All other materials are to be of the best of their respective kinds.

Estimate for Sinking and Lining a Well 80 feet deep and 4 feet 6 inches in Diameter.

Cubic yards.		At per cubic yard.		£ s. d.		
		s.	*d.*	£	*s.*	*d.*
	Excavation.					
9¾	Alluvium, sand and gravel	1	9		17	1
23	Soft marl	4	6	5	3	6
52¼	Hard marl	6	6	16	19	7
5¾	Sandstone	10	0	2	17	6
	Brickwork.					
32	Brickwork in cement ...	40	0	64	0	0
	Concrete.					
11¾	Portland cement concrete	16	0	9	8	0
	Miscellaneous.					
	Elm windlass, frame, bucket and rope complete			3	1	6
	Wooden cover-doors ...			0	6	6
	Removing excavation to spoil			4	10	0
	Lifting water			5	0	0
				112	3	8

Lead Poisoning.

Constant reference has been made in the course of these pages to the solvent properties which water sometimes possesses with regard to lead. This subject was forced upon public attention a few years ago, on account of its serious consequences at Sheffield and other northern towns, where the water is very soft in character, and is principally derived from moorland. These alarming outbreaks led to much careful investigation both as regards the peculiar constituents of such waters as are most liable to become polluted with salts of lead, as well as the circumstances which tend to facilitate such pollution.

Lead is a cumulative poison, and if water containing the most minute quantities is constantly employed for dietetic purposes, lead poisoning (plumbism or saturnism) must eventually supervene. The absorption of lead into the system constitutes a predisposing cause of many diseases, and there is a liability of symptomatic treatment, the true origin of the disorder being undiscovered. In fact, the theory has been propounded that nearly all cases of gout, Bright's disease (nephritis), and many other diseases, might be traced to lead poisoning.

There is perhaps no subject upon which more diversity of opinion exists amongst scientists than that which relates, firstly to the essential characteristics of the water itself, and secondly, to the most favourable conditions under which water will attack lead. With regard to the activity of the water there are several theories :—

1. Presence of acidity in the water.

2. Insufficiency of dissolved silica.

3. Absence of a sufficient proportion of dissolved carbonic acid (CO_2).

4. Deficiency of salts, especially of phosphates, carbonates, and sulphates.

5. Presence of sewage matter, especially of nitrates and nitrites.

1. *Presence of Acidity in the Water.*

There is no doubt that certain waters, especially from moorland sources, possess a distinct acidity, and that their lead-dissolving properties are directly proportional to their degree of acidity. This was shown to be the case by the experiments conducted by Dr. Sinclair White in connection with the Sheffield outbreak, and communicated in a paper read at the meeting of the British Medical Association, in Leeds, in August, 1889.

The nature of this acid is, we believe, yet a mystery. One opinion is that it consists of sulphuric acid, derived from the oxidation of iron pyrites; another opinion is that the acid is of vegetable origin, and is due to the decomposition of vegetable matter (*e.g.* peat), possibly due to the action of bacteria.

The remedies proposed in this case are—

(*a*) Contact with fragments of limestone.

(*b*) Admixture of a proportion of milk of lime.

(*c*) Admixture of a proportion of carbonate of soda.

Contact with fragments of limestone, in addition to the admixture with the water of a certain quantity of quicklime, has been adopted with satisfactory results at Keighley. As the acidity of the water varies from time to time, the quantity of alkali necessary should be periodically determined by analysis, and added to the water in the form of powder or as milk of lime. This is the more important as it has been proved that an excess of lime increases the activity of water towards lead. Dr. Tidy is reported to have said that the beneficial action of limestone was due simply to the silica which it contained. Fragments of limestone become coated after a few weeks and require renewal; brushing is said to be effective.

The admixture of a proportion of carbonate of soda with the water has been strongly advocated by Dr. Percy F. Frankland (Sanitary Institute Congress, Brighton, 1890). The quantity to be used must be determined from time to time by analysis, 5 parts of carbonate of soda to 100,000 parts of water being an extreme case. This process has been adopted at Wakefield.

2. *Insufficiency of Dissolved Silica.*

This opinion has been strongly advocated by Dr. Tidy and other eminent chemists, as the result of observations which appeared to show that the activity of a water towards lead was destroyed when it contained upwards of half a grain of silica per gallon. It would seem, however, that this is not invariably the case, from the following instances :—

(*a*) Water from the Punch Bowl, Hindhead, which

contains 0·831 grains of silica per gallon, is said to act vigorously on lead.

(*b*) Experiments made by Professor Williams, of Sheffield, showed that silica added in definite quantities to an acid water did not diminish its solvent action upon lead.

(*c*) The High Level water at Sheffield acted vigorously upon lead whilst containing a larger proportion of silica than the Low Level water, which acted slightly or not at all upon lead.

Besides this, Dr. Sinclair White (1889) is of opinion that "the amount of silica which moorland water will take up from flints, even after long contact, is very small, and in practice it would seem to be exceedingly difficult, if not impossible, to silicate by means of ordinary flints some of these waters to the extent of containing half a grain per gallon. Mr. A. H. Allen has never succeeded in adding more than a quarter of a grain per gallon in this way, and Professor Percy Frankland's experience is of a similar character."

3. *Absence of a Sufficient Proportion of Dissolved Carbonic Acid* (CO_2).

This is tantamount to saying that a water should possess a certain degree of temporary hardness, and this theory is borne out by Dr. Percy Frankland, who states that such waters "may be generally considered above suspicion."

4. *Deficiency of Salts, especially of Phosphates, Carbonates, and Sulphates.*

In other words, soft waters are dangerous, and perhaps this is the most popular opinion upon the subject. It must be remembered, however, as already stated, that an *excess* of lime in a water tends to increase its activity towards lead; and that hard waters, which derive their salts from sewage-polluted sources (nitrates, nitrites, etc.), are especially active towards lead.

o

5. *Presence of Sewage-matter, especially of Nitrates, and Nitrites.*

Although this condition of a water undoubtedly renders it more active towards lead, it is almost needless to say that such pollution is not an indispensable cause.

It would appear that much time has been wasted in attempting to trace the property possessed by certain waters of attacking lead to one final cause, instead of admitting that many causes may be at work at the same time, either separately or conjointly.

With regard to the most favourable conditions under which water will attack lead, there is nearly as much variety of opinion.

A frequently stated dogma is that new lead only is to be feared, old lead becoming protected by a coating of carbonate of lead which forms upon its surface. Dr. Frankland has, however, shown that some waters act more and more upon pipes from day to day. It was originally believed that carbonate and sulphate of lead were insoluble in water; this was due to an error in the method adopted for analysis, which assumed that sulphide of lead was insoluble in water charged with sulphuretted hydrogen. Again, it has been shown that the coating of carbonate of lime and lead, which forms on a lead surface exposed to an active water, is pervious and does not act as a protection. This coating also is very liable to scale off as a result of vibration, *e.g.* in pipes, the opening and closing of taps, the passage of vehicles, change of temperature, etc.

The influence of pressure upon the action of the water has also led to much diversity of opinion. Dr. Sinclair White states, as the result of experiment, that, "Other things being equal, the greater the pressure under which the water is stored the greater amount of lead is taken up. This influence is considerable, but no amount of pressure will, of itself, render a harmless water active towards lead."

There is more agreement upon the influence of temperature,

and Dr. Sinclair White's statement that, "Other things being equal, an increase in the temperature of the water increases its lead-dissolving power," may be taken as the general opinion. This point, taken in connection with the remarks made above upon "temporary hardness," is interesting.

It seems to be generally accepted that lead surfaces exposed alternately to the action of air and water are more liable to be attacked; also that water becomes more active by becoming charged with air.

Where the water is acidulated its action is much increased when it can bring into circuit with the lead surfaces of iron, copper, zinc, brass, etc.; in such cases an electric current appears to be set up, the lead being the soluble electrode.

The importance of investigating the action of a proposed water-supply upon lead must be impressed upon the student. It must be borne in mind, in conducting this investigation, that the quality of a water from the same source varies considerably in this respect, and that all the conditions capable of affecting the question must therefore be carefully ascertained.

When, however, the active property of the water towards lead has been discovered subsequently to the construction of the works, in addition to the best practical measures being taken by the authority to counteract the solvent property before the water leaves their reservoir, information (in the form of leaflets) should be distributed amongst the consumers, so as to enable them to take such protective measures as will minimize the danger. The following recommendations have been suggested by **Dr. Frankland** :—

(1) That no water should be collected for drinking purposes until after the tap has been allowed to run for such a length of time as will presumably clear the service-pipe, and that drinking or cooking water may, therefore, be advantageously collected immediately after a considerable quantity of water has been drawn for other domestic purposes.

(2) That the filtration of the water through any form of animal charcoal filter, practically guarantees its absolute freedom from lead.

(3) That hot water acts more powerfully on lead than cold, and that, therefore, metal tea-pots and other soldered vessels for holding hot water should be avoided as much as possible.

It is an interesting fact, not yet satisfactorily explained, that filters composed of "animal charcoal" have the property not only of removing lead after it has been dissolved, but of removing from an active water its property of dissolving lead, and that this property is continuous.

Another point to be remembered is that the lead from which pipes are usually constructed is not chemically pure. They generally consist of two-thirds of new and one-third of old lead, the latter having been already used, and containing tin, zinc, antimony, and other metals, which facilitate the formation of electric currents.

CHAPTER XXVI.

PUBLIC INQUIRIES. CONCLUSION.

In the preliminary chapter it was stated that the money for carrying out waterworks by Local Authorities is usually obtained upon loan after a Local Government Board inquiry.

The Public Health Act, 1875, sec. 293, empowers the Local Government Board to cause to be made from time to time, "such inquiries as are directed by this Act, and such inquiries as they see fit in relation to any matters concerning the public health in any place, or any matter with respect to which their sanction, approval, or consent is required by this Act."

The principal matters connected with the supply of water to rural districts upon which public inquiries are held, are—

1. The borrowing of money beyond a certain amount.

2. The carrying of water-mains without the district.

3. The purchase of lands otherwise than by agreement.

4. The formation of a special drainage district for purposes of water supply.

5. The construction of large reservoirs.

The Public Health Act, 1875, sec. 295, states that "all orders made by the Local Government Board in pursuance of this Act shall be binding and conclusive in respect of the matters to which they refer, and shall be published in such manner as the Board may direct."

Sec. 296 empowers inspectors of the Local Government Board "for the purposes of any inquiry directed by the Board (to) have in relation to witnesses and their examination, the production of papers and accounts, and the inspection of places and matters required to be inspected, similar powers to those which Poor Law Inspectors have under the Acts relating to the relief of the poor for the purposes of those Acts."

1. THE BORROWING OF MONEY.

By secs. 233, 234, Local Authorities are empowered, with the sanction of the Local Government Board, to borrow money for "permanent works" (*e.g.* waterworks), and for such purpose they may "mortgage to the person by or on behalf of whom such sums are advanced, any funds or rates out of which they are authorized to defray expenses incurred by them in the execution of this Act." In the case of a Rural Authority the cost of providing a supply of water to any contributory place within the district (*e.g.* parish) is by sec. 229, a special expense, and only the rate or rates out of which such expenses are payable may be mortgaged for that purpose.

The total amount of the loans outstanding is not at any time to exceed the "assessable" value for two years of the premises assessable within the district in respect of which such money may be borrowed. By "assessable" must be understood "rateable" in this connection.

Where the sum proposed to be borrowed, together with the balances of outstanding loans (if any), would exceed the assessable value for one year of such premises, the Local Government Board shall not give their sanction to such loan until one of their inspectors has held a local inquiry and reported to the said Board.

The loan is usually obtained from the "Public Works

Loan Commissioners," on the recommendation of the Local
Government Board. It must be remembered that the
Local Government Board can only recommend, and cannot
compel.

2. THE CARRYING OF WATER-MAINS WITHOUT THE DISTRICT.

Sec. 32 requires a Local Authority, before commencing the
construction or extension (sec. 54) of any water-main without the district, to give three months' notice by advertise-
ment in one or more of the local newspapers circulated
within the district where the work is to be made. Such
notice is to describe the nature of the intended work, and
shall state the intended termini thereof, and the names of the
parishes, and the turnpike roads and streets, and other lands
(if any), through, across, under, or on which the work is to
be made, and shall name a place where a plan of the intended work is open for inspection at all reasonable hours;
and a copy of such notice shall be served on the owners or
reputed owners, lessees or reputed lessees, and occupiers of
the said lands, and on the overseers of such parishes, and on
the trustees, surveyors of highways, or other persons having
the care of such roads or streets.

Sec. 33 is to the effect that if any objection is raised
against the intended works the said works must not be
commenced without the sanction of the Local Government
Board.

Sec. 34. "The Local Government Board may, on application of the Local Authority, appoint an inspector to make
inquiry on the spot into the propriety of the intended work
and into the objections thereto, and to report to the Local
Government Board on the matters with respect to which
such inquiry was directed, and on receiving the report of
such inspector the Local Government Board may make an
order disallowing or allowing, with such modifications (if
any) as they may deem necessary, the intended work."

Sec. 285 provides that "any Local Authority may, with the consent of the Local Authority of any adjoining district, execute and do in such adjoining district all or any of such works and things as they may execute and do within their own district, and on such terms as to payment or otherwise as may be agreed on between them and the Local Authority of the adjoining district."

An important issue has arisen in connection with the two last clauses, which is clearly set forth in the following extract from the *Justice of the Peace* for May 19, 1894, which applies equally to water-mains (sec. 54):—

"Query: (2) If one authority desires to carry a sewer through land of an adjoining authority, will it be sufficient to comply with sec. 32, or will it be necessary to obtain the consent mentioned in sec. 285?

"(3) If such consent is necessary, can it be arbitrarily withheld by the adjoining authority?

"Answer: (2) We think the consent of the adjoining authority must be obtained. Such consent is a condition precedent to the works being undertaken.

"(3) If the consent is withheld, there is no power of compelling the consent to be given. It is, therefore, immaterial whether or not the consent is arbitrarily withheld, as such consent is necessary before undertaking the works."

3. The Purchase of Land otherwise than by Agreement.

A Local Authority, after having complied with the requirements of sec. 176 (sub-sec. 2), with regard to publication and the service of notices, may petition the Local Government Board to permit them to put into force the Land Clauses Consolidation Acts. After receiving such petition, and being satisfied that the necessary formalities as to publication and service of notices have been complied with, the Local Government Board may either dismiss the petition or institute an inquiry. After such inquiry the Local Governmen ·Board may grant the petition, with such modifications or conditions that the Board may think fit.

4. The Formation of Special Drainage Districts.

As stated in Chapter I., the Local Government Board are rarely in favour of this step for purposes of water supply alone.

5. The Construction of Large Reservoirs.

This refers only to reservoirs other than service-reservoirs or tanks which will hold not more than 100,000 gallons, and therefore rarely applies to questions of rural water-supply. The Local Authority is required to properly advertise the proposed work in the local newspapers, and if any person who would be affected by the intended work lodges an objection, the Local Authority may appeal to the Local Government Board. After receiving such appeal the Local Government Board may institute an inquiry, after which they may make an order disallowing or allowing, with such modifications (if any), as they may deem necessary, the intended work.

Where a Local Authority make default (*inter alia*), in providing their district with a supply of water (sec. 299), and complaint is made to the Local Government Board, the Local Government Board, "if satisfied, after due inquiry, that the authority has been guilty of the alleged default, shall make an order limiting a time for the performance of their duty in the matter of such complaint. If such duty is not performed by the time limited in the order, such order may be enforced by writ of mandamus, or the Local Government Board may appoint some person to perform such duty." Similar powers are now given by the Local Government Act, 1894 (sec. 16), to County Councils upon the complaint of Parish Councils.

Before holding an inquiry, the Local Government Board require to be supplied with certain information in the form of statistics, plans, estimates, and details. The requirements of the Local Government Board as to plans have already been dealt with in Chap. XI. The remaining information has to be supplied upon forms issued for the purpose by the Local Government Board, which are filled up by the clerk or engineer, according to the nature of the queries.

The appended form has to be filled up by the engineer of the proposed works—

> Works of Water Supply.
>
> Estimates and Details.
>
> Name of Sanitary Authority ———.

In the case of a Rural Sanitary Authority, name of contributory place for which the works are required ———.

Estimate for Cast-iron Mains.

Name of street or road.	Internal diameter of pipes in inches.	Weight per yard.	Total length in yards.	Total weight.	To carry head in feet vertical.	To carry lbs. pressure on the inch.	Price per cwt.	Total.
		cwts. lbs.		cwts. lbs.				£ s. d.

ESTIMATE FOR FITTINGS.

Description of Work.	Nos.	Internal diameter in inches of valves, etc.	Weight of each in cwts. and lbs.	Price for cash complete.	Total.	Remarks.
					£ s. d.	
Fire-cocks and hydrants						
Casings to fire-cocks and hydrants complete						
Sluice-valves						
Casings to sluice-valves complete						

N.B.—Describe the sluice-valves, fire-cocks and hydrants, and state if or not the valves, or any portion of them, are to be of gun-metal.

Describe the casings and street fittings in detail.

Furnish diagrams if they have been prepared.

DETAILS OF THE WORKS.

HEADINGS FOR A DETAILED DESCRIPTION OF THE WORKS TO BE FURNISHED BY THE ENGINEER.

Pumping Works.

1. Sort of engine proposed.
2. Estimated power of engine.
3. Estimated weight of coal per hour, per horse-power.
4. Volume of water to be lifted.
5. Head to which water is to be lifted.

6. Internal diameter of rising-main, in inches.

7. Calculated velocity of water in feet per second, through the rising-main.

Impounding and Storage Reservoir.

1. Area of gathering ground, in acres.
2. Average annual depth of rainfall.
3. Surface area of reservoir when full, in acres.
4. Greatest depth of water in the reservoir when full, in feet.
5. Total volume when full, in gallons.
6. Length of bye-wash.
7. Height of the embankment above top water-level, and top width and thickness of the puddle-wall at bottom and at top.

Covered Service Reservoir.

1. Area of reservoir in square yards.
2. Depth of water in reservoir when full, in feet.
3. Volume of water in reservoir when full, in gallons.

N.B.—Describe in writing the proposed mode of construction of covered reservoirs; as, also, how the reservoir is to be ventilated, lighted, and worked.

It is not desirable to make the pumping main a supply main also, nor should the velocity through any pumping main exceed 2 feet per second.

The whole of the cast-iron pipes and other castings must be varnished.

There should not be less than 3 feet in length of bye-wash for each 100 acres of gathering ground.

Add any additional details necessary to a full explanation of the proposed works.

Plans and sections, or tracings of them, together with details, must be furnished.

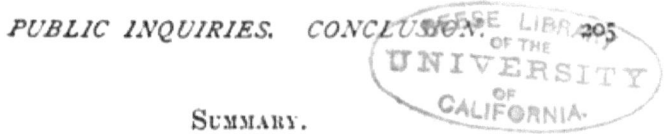

SUMMARY.

Description of work.	Cost.	Total cost.	Remarks.
	£ s. d.	£ s. d.	

Date——   Signed——

N.B.—This form should be signed by the engineer of the proposed works.

At the inquiry the engineer is called upon to describe and explain the proposed works, and is frequently subjected to a sharp cross-examination upon the details. It is therefore necessary that he should be fully prepared upon every point, and have his notes in such a form as to be ready for reference at a moment's notice. A useful method is to use a note-book with an alphabetical margin, and to arrange the various matters alphabetically. This practice will be found simple and rapid. Another useful practice is to lay down the mains and branches upon a 1-inch Ordnance map indicating the positions of reservoirs, tanks, sluice-valves, hydrants, etc., and distinguishing by means of different colours the ownership of all lands built upon, passed through, or in any way affected by the intended works.

In conclusion we would urge upon the student the necessity for careful thought and preparation before advancing any scheme for water supply. The details comprehended in the profession of a waterworks engineer are innumerable, but upon the full appreciation of these details will depend his success or failure. A small scheme requires, in its way, as much preparation as a large one, and an error in calculation, which in the latter would be insignificant, might be the ruin of the former.

The habits of careful investigation, unerring accuracy, and steady perseverance, combined with a thorough practical knowledge, are qualifications which alone will lead to the execution of those works with which the engineer may afterwards be proud to hear his name associated.

INDEX.

THE END.

PRINTED BY WILLIAM CLOWES AND SONS, LIMITED, LONDON AND BECCLES.

THE
WATER SUPPLY OF TOWNS,
AND THE
CONSTRUCTION OF WATERWORKS:
A Practical Treatise for the use of Engineers and Students of Engineering

BY

W. K. BURTON, A.M.INST.C.E.,
Professor of Sanitary Engineering in the Imperial University, Tokyo, Japan.

TO WHICH IS APPENDED

A PAPER ON THE EFFECTS OF EARTHQUAKES ON WATERWORKS
By Professor JOHN MILNE, F.R.S.

WITH NUMEROUS PLATES AND OTHER ILLUSTRATIONS.

SUMMARY OF CONTENTS.

OPINIONS OF THE PRESS.

"The chapter upon filtration of water is very complete, and the details of construction well illustrated. . . . The work should be specially valuable to civil engineers engaged in work in Japan, but the interest is by no means confined to that locality."—*Engineer.*

"It is with great pleasure that we chronicle an addition to the literature of this important branch of engineering, and we congratulate the author upon the practical common sense shown in the preparation of this work. . . . The plates and diagrams have evidently been pre-pared with great care, and cannot fail to be of great assistance to the student."—*Builder.*

"The whole art of waterworks construction is dealt with in a clear and comprehensive fashion in this handsome volume. . . . Mr. Burton's practical treatise shows in all its sections the fruit of independent study and individual experience. It is largely based upon his own practice in the branch of engineering of which it treats."—*Saturday Review.*

LONDON: CROSBY LOCKWOOD & SON, 7, Stationers' Hall Court, E.C.

A
COMPREHENSIVE TREATISE
ON THE
Water Supply of Cities & Towns

BY

WILLIAM HUMBER, A.M.Inst.C.E., & M.Inst.M.E.,

AUTHOR OF

"CAST AND WROUGHT-IRON BRIDGE CONSTRUCTION," Etc.

Illustrated with 50 Double Plates, 1 Single Plate, Coloured Frontispiece, and upwards of 250 Woodcuts, and containing 400 pages of Text.

SUMMARY OF CONTENTS.

OPINIONS OF THE PRESS.

"The most systematic and valuable work upon water supply hitherto produced in English, or in any other language. . . . Mr. Humber's work is characterized almost throughout by an exhaustiveness much more distinctive of French and German than of English technical treatises."—*Engineer.*

"We can congratulate Mr. Humber on having been able to give so large an amount of information on a subject so important as the water supply of cities and towns. The plates, fifty in number, are mostly drawings of executed works, and alone would have commanded the attention of every engineer whose practice may lie in this branch of the profession."—*Builder.*

LONDON: CROSBY LOCKWOOD & SON, 7, Stationers' Hall Court, E.C.

Second Edition, Crown 8vo, 320 pages, with Illustrations, 7s. 6d., cloth.

WATER ENGINEERING
A PRACTICAL TREATISE
ON THE MEASUREMENT, STORAGE, CONVEYANCE, AND UTILIZATION OF
WATER FOR THE SUPPLY OF TOWNS, FOR MILL POWER, AND FOR
OTHER PURPOSES.
By CHARLES SLAGG,
Water and Drainage Engineer, Associate Member of the Institution of Civil Engineers, Author
of "Sanitary Work in the Smaller Towns, and in Villages."

SUMMARY OF CONTENTS.

EMBANKMENTS OF WATERWORKS RESERVOIRS—QUANTITY OF WATER TO BE STORED—APPURTENANCES OF RESERVOIRS—DISCHARGE OF WATER FROM THE RESERVOIR—APPROXIMATE COST OF A STORAGE RESERVOIR—CONCRETE FOR EMBANKMENTS AND DAMS—STREAM GAUGES—RAINFALL—AREAS OF RIVER-BASINS — CONDUITS — TUNNELS—SERVICE RESERVOIRS — PRESSURE AND ITS EFFECTS IN PIPES — AQUEDUCTS — RIVERS AND WATERCOURSES — COMPENSATION TO

MILLS—OF WATER-POWER IN GENERAL—WATER WHEELS—CORN MILLS—WORK DONE BY WATER WHEELS—TURBINES—DOMESTIC WATER SUPPLY — SERVICE RESERVOIRS — DISTRIBUTION OF WATER—PUMPING-MAINS AND ENGINES—FLOODS—STORAGE OF FLOOD WATERS—RELIEF OF LAND FROM FLOODS—REGULATION OF FLOOD WATERS — RIVER CONSERVANCY — COUNTRY BOARDS AND WATERSHED AREAS.

"As a small practical treatise on the water supply of towns, and on some applications of water-power, the work is in many respects excellent."—*Engineering.*

"The author has collated the results deduced from the experiments of the most eminent authorities, and has presented them in a compact and practical form, accompanied by very clear and detailed explanations. . . . The application of water as a motive power is treated very carefully and exhaustively."—*Builder.*

LONDON: CROSBY LOCKWOOD & SON, 7, Stationers' Hall Court, E.C.

Fcap. 8vo, nearly 400 pages, price 7s. 6d., cloth.

THE HEALTH OFFICER'S POCKET-BOOK
A GUIDE TO SANITARY PRACTICE AND LAW FOR MEDICAL OFFICERS OF HEALTH, SANITARY INSPECTORS, MEMBERS OF SANITARY AUTHORITIES, ETC.
By EDWARD F. WILLOUGHBY, M.D. (Lond.), D.P.H. (Lond. & Camb.),
Author of "Hygiene and Public Health," &c.

SUMMARY OF CONTENTS.

PART I.—PRACTICAL HYGIENE.

CHAP.
I.—MATHEMATICAL PRACTICE.
II.—METEOROLOGICAL PRACTICE.
III.—DEMOGRAPHY AND STATISTICS.
IV.—ENGINEERING MEMORANDA.
V.—SANITARY PRACTICE.
VI.—WATER.
VII.—DIETETICS.
VIII.—SCAVENGING.

PART II.—SANITARY LAW.

ENUMERATION OF CLAUSES IN THE GREATER ACTS.
ABSTRACTS OF ACTS IN PARLIAMENT.
MINOR ACTS OF PARLIAMENT.
LIST OF MODEL BYELAWS OF LOCAL GOVERNMENT BOARD.
FORMS OF NOTICES, ORDERS, &c.

"It is a mine of condensed information of a pertinent and useful kind on the various subjects of which it treats. The matter seems to have been carefully compiled and arranged for facility of reference, and it is well illustrated by diagrams and woodcuts. The different subjects are succinctly but fully and scientifically dealt with."—*The Lancet.*

"We feel quite safe in recommending all those engaged in practical sanitary work to furnish themselves with a copy for reference. We can also recommend it for study to those going in for the various examinations in public health, as it contains a large amount of valuable information on subjects which frequently appear in the examination papers, and which are not generally found in works on hygiene and public health."—*Sanitary Journal.*

LONDON: CROSBY LOCKWOOD & SON, 7, Stationers' Hall Court, E.C.

7, STATIONERS' HALL COURT, LONDON, E.C.

September, 1895.

A

CATALOGUE OF BOOKS

INCLUDING NEW AND STANDARD WORKS IN
ENGINEERING: CIVIL, MECHANICAL, AND MARINE;
ELECTRICITY AND ELECTRICAL ENGINEERING;
MINING, METALLURGY; ARCHITECTURE,
BUILDING, INDUSTRIAL AND DECORATIVE ARTS;
SCIENCE, TRADE AND MANUFACTURES;
AGRICULTURE, FARMING, GARDENING;
AUCTIONEERING, VALUING AND ESTATE AGENCY;
LAW AND MISCELLANEOUS.

PUBLISHED BY

CROSBY LOCKWOOD & SON.

MECHANICAL ENGINEERING, etc.

D. K. Clark's Pocket-Book for Mechanical Engineers.

THE MECHANICAL ENGINEER'S POCKET-BOOK OF TABLES, FORMULÆ, RULES AND DATA. A Handy Book of Reference for Daily Use in Engineering Practice. By D. KINNEAR CLARK, M.Inst.C.E., Author of "Railway Machinery," "Tramways," &c. Second Edition, Revised and Enlarged. Small 8vo, 700 pages, 9s. bound in flexible leather covers, with rounded corners and gilt edges.

SUMMARY OF CONTENTS.

MATHEMATICAL TABLES.—MEASUREMENT OF SURFACES AND SOLIDS.—ENGLISH WEIGHTS AND MEASURES.—FRENCH METRIC WEIGHTS AND MEASURES.—FOREIGN WEIGHTS AND MEASURES.—MONEYS.—SPECIFIC GRAVITY WEIGHT AND VOLUME.—MANUFACTURED METALS.—STEEL PIPES.—BOLTS AND NUTS.—SUNDRY ARTICLES IN WROUGHT AND CAST IRON, COPPER, BRASS, LEAD, TIN, ZINC.—STRENGTH OF MATERIALS.—STRENGTH OF TIMBER.—STRENGTH OF CAST IRON.—STRENGTH OF WROUGHT IRON.—STRENGTH OF STEEL.—TENSILE STRENGTH OF COPPER, LEAD, ETC.—RESISTANCE OF STONES AND OTHER BUILDING MATERIALS.—RIVETED JOINTS IN BOILER PLATES.—BOILER SHELLS.—WIRE ROPES AND HEMP ROPES.—CHAINS AND CHAIN CABLES.—FRAMING.—HARDNESS OF METALS, ALLOYS AND STONES.—LABOUR OF ANIMALS.—MECHANICAL PRINCIPLES.—GRAVITY AND FALL OF BODIES.—ACCELERATING AND RETARDING FORCES.—MILL GEARING, SHAFTING, ETC.—TRANSMISSION OF MOTIVE POWER.—HEAT.—COMBUSTION: FUELS.—WARMING, VENTILATION, COOKING STOVES.—STEAM.—STEAM ENGINES AND BOILERS.—RAILWAYS—TRAMWAYS.—STEAM SHIPS.—PUMPING STEAM ENGINES AND PUMPS.—COAL GAS, GAS ENGINES, ETC.—AIR IN MOTION.—COMPRESSED AIR.—HOT AIR ENGINES.—WATER POWER.—SPEED OF CUTTING TOOLS.—COLOURS.—ELECTRICAL ENGINEERING.

*** OPINIONS OF THE PRESS.

"Mr. Clark manifests what is an innate perception of what is likely to be useful in a pocket-book, and he is really unrivalled in the art of condensation. Very frequently we find the information on a given subject is supplied by giving a summary description of an experiment, and a statement of the results obtained. There is a very excellent steam table, occupying five and-a-half pages; and there are rules given for several calculations, which rules cannot be found in other pocket-books, as, for example, that on page 407, for getting at the quantity of water in the shape of priming in any known weight of steam. It is very difficult to hit upon any mechanical engineering subject concerning which this work supplies no information, and the excellent index at the end adds to its utility. In one word, it is an exceedingly handy and efficient tool, possessed of which the engineer will be saved many a wearisome calculation, or yet more wearisome hunt through various text-books and treatises, and, as such, we can heartily recommend it to our readers, who must not run away with the idea that 'Mr. Clark's Pocket-book is only Molesworth in another form. On the contrary, each contains what is not to be found in the other; and Mr. Clark takes more room and deals at more length with many subjects than Molesworth possibly could."

The Engineer.

"It would be found difficult to compress more matter within a similar compass, or produce a book of 650 pages which should be more compact or convenient for pocket reference. . . Will be appreciated by mechanical engineers of all classes."—*Practical Engineer.*

"Just the kind of work that practical men require to have near to them."—*English Mechanic.*

B

MR. HUTTON'S PRACTICAL HANDBOOKS.

Handbook for Works' Managers.

THE WORKS' MANAGER'S HANDBOOK OF MODERN RULES, TABLES, AND DATA. For Engineers, Millwrights, and Boiler Makers; Tool Makers, Machinists, and Metal Workers; Iron and Brass Founders, &c. By W. S. HUTTON, Civil and Mechanical Engineer, Author of "The Practical Engineer's Handbook." Fifth Edition, carefully Revised, with Additions. In One handsome Volume, medium 8vo, price 15s. strongly bound.   [*Just published.*

☞ *The Author having compiled Rules and Data for his own use in a great variety of modern engineering work, and having found his notes extremely useful, decided to publish them—revised to date—believing that a practical work, suited to the* DAILY *REQUIREMENTS OF MODERN ENGINEERS, would be favourably received.*

In the Fourth Edition the First Section has been re-written and improved by the addition of numerous Illustrations and new matter relating to STEAM ENGINES *and* GAS ENGINES. *The Second Section has been enlarged and Illustrated, and throughout the book a great number of emendations and alterations have been made, with the object of rendering the* **book** *more generally useful.*

. OPINIONS OF THE PRESS.

" The author treats every subject from the point of view of one who has collected workshop notes for application in workshop practice, rather than from the theoretical or literary aspect. The volume contains a great deal of that kind of information which is gained only by practical experience, and is seldom written in books."—*Engineer.*

" The volume is an exceedingly useful one, brimful with engineers' notes, memoranda, and rules, and well worthy of being on every mechanical engineer's bookshelf."—*Mechanical World.*

" The information is precisely that likely to be required in practice. . . . The work forms a desirable addition to the library not only of the works' manager, but of anyone connected with general engineering."—*Mining Journal.*

" A formidable mass of facts and figures, readily accessible through an elaborate index . . . Such a volume will be found absolutely necessary as a book of reference in all sorts of 'works' connected with the metal trades."—*Ryland's Iron Trades Circular.*

" Brimful of useful information, stated in a concise form, Mr. Hutton's books have met a pressing want among engineers. The book must prove extremely useful to every practical man possessing **a copy.**"—*Practical Engineer.*

New Manual for Practical Engineers.

THE PRACTICAL ENGINEER'S HAND-BOOK. Comprising a Treatise on Modern Engines and Boilers: Marine, Locomotive and Stationary. And containing a large collection of Rules and Practical Data relating to recent Practice in Designing and Constructing all kinds of Engines, Boilers, and other Engineering work. The whole constituting a comprehensive Key to the Board of Trade and other Examinations for Certificates of Competency in Modern Mechanical Engineering. By WALTER S. HUTTON, Civil and Mechanical Engineer, Author of " The Works' Manager's Handbook for Engineers," &c. With upwards of 370 Illustrations. Fourth Edition, Revised, with Additions. Medium 8vo, nearly 500 pp., price 18s. Strongly bound.

☞ *This work is designed as a companion to the Author's* "WORKS' MANAGER'S HAND-BOOK." *It possesses many new and original features, and contains, like its predecessor, a quantity of matter not originally intended for publication, but collected by the author for his own use* **in the** *construction of a great variety of* MODERN ENGINEERING WORK.

The information is given in a condensed and **concise form, and is illustrated by** *upwards of 370 Woodcuts; and comprises* **a quantity of tabulated matter of great** *value to all engaged in designing, constructing, or* **estimating for** ENGINES, BOILERS, *and* OTHER ENGINEERING WORK.

. OPINIONS OF THE PRESS.

" We have kept it at hand for several weeks, referring to it as occasion arose, and we have not on a single occasion consulted its pages without finding the information of which we were in quest. —*Athenæum.*

" A thoroughly good practical handbook, which no engineer can go through without learning something that will be of service to him."—*Marine Engineer.*

" An excellent book of reference for engineers, and a valuable text-book for students of engineering."—*Scotsman.*

" This valuable manual embodies the results and experience of the leading authorities on mechanical engineering."—*Building News.*

" The author has collected together a surprising quantity of rules and practical data, and has shown much judgment in the selections he has made. . . . There is no doubt that this book is one of the most useful of its kind published, and will be a very popular compendium."—*Engineer.*

" A mass of information, set down in simple language, and in such a form that it can be easily referred to at any time. The matter is uniformly good and well chosen and is greatly elucidated by the illustrations. The book will find its way on to most engineers' shelves, where it will rank as one of the most useful books of reference."—*Practical Engineer.*

" Full of useful information and should be found on the office shelf of all **practical engineer**." —*English Mechanic.*

MR. HUTTON'S PRACTICAL HANDBOOKS—continued.

Practical Treatise on Modern Steam-Boilers.

STEAM-BOILER CONSTRUCTION. A Practical Handbook for Engineers, Boiler-Makers, and Steam Users. Containing a large Collection of Rules and Data relating to Recent Practice in the Design, Construction, and Working of all Kinds of Stationary, Locomotive, and Marine Steam-Boilers. By WALTER S. HUTTON, Civil and Mechanical Engineer, Author of "The Works' Manager's Handbook," "The Practical Engineer's Handbook," &c. With upwards of 300 Illustrations. Second Edition. Medium 8vo, 18s. cloth.

☞ *This work is issued in continuation of the Series of Handbooks written by the Author, viz:—"THE WORKS' MANAGER'S HANDBOOK" and "THE PRACTICAL ENGINEER'S HANDBOOK," which are so highly appreciated by Engineers for the practical nature of their information; and is consequently written in the same style as those works.*

The Author believes that the concentration, in a convenient form for easy reference, of such a large amount of thoroughly practical information on Steam-Boilers, will be of considerable service to those for whom it is intended, and he trusts the book may be deemed worthy of as favourable a reception as has been accorded to its predecessors.

** OPINIONS OF THE PRESS.

"Every detail, both in boiler design and management, is clearly laid before the reader. The volume shows that boiler construction has been reduced to the condition of one of the most exact sciences; and such a book is of the utmost value to the *fin de siècle* Engineer and Works Manager."—*Marine Engineer.*

"There has long been room for a modern handbook on steam boilers; there is not that room now, because Mr. Hutton has filled it. It is a thoroughly practical book for those who are occupied in the construction, design, selection, or use of boilers."—*Engineer.*

"The book is of so important and comprehensive a character that it must find its way into the libraries of everyone interested in boiler using or boiler manufacture if they wish to be thoroughly informed. We strongly recommend the book for the intrinsic value of its contents."—*Machinery Market.*

"The value of this book can hardly be over-estimated. The author's rules, formulæ, &c., are all very fresh, and it is impossible to turn to the work and not find what you want. No practical engineer should be without it."—*Colliery Guardian.*

Hutton's "Modernised Templeton."

THE PRACTICAL MECHANICS' WORKSHOP COMPANION. Comprising a great variety of the most useful Rules and Formulæ in Mechanical Science, with numerous Tables of Practical Data and Calculated Results for Facilitating Mechanical Operations. By WILLIAM TEMPLETON, Author of "The Engineer's Practical Assistant," &c. &c. Seventeenth Edition, Revised, Modernised, and considerably Enlarged by WALTER S. HUTTON, C.E., Author of "The Works' Manager's Handbook," "The Practical Engineer's Handbook," &c. Fcap. 8vo, nearly 500 pp., with 8 Plates and upwards of 250 Illustrative Diagrams. 6s., strongly bound for workshop or pocket wear and tear.

** OPINIONS OF THE PRESS.

"In its modernised form Hutton's 'Templeton' should have a wide sale, for it contains much valuable information which the mechanic will often find of use, and not a few tables and notes which he might look for in vain in other works. This modernised edition will be appreciated by all who have learned to value the original editions of 'Templeton.'"—*English Mechanic.*

"It has met with great success in the engineering workshop, as we can testify; and there are a great many men who, in a great measure, owe their rise in life to this little book."—*Building News.*

"This familiar text-book—well known to all mechanics and engineers—is of essential service to the every-day requirements of engineers, millwrights, and the various trades connected with engineering and building. The new modernised edition is worth its weight in gold."—*Building News.* (Second Notice.)

"This well-known and largely used book contains information, brought up to date, of the sort so useful to the foreman and draughtsman. So much fresh information has been introduced as to constitute it practically a new book. It will be largely used in the office and workshop."—*Mechanical World.*

"The publishers wisely entrusted the task of revision of this popular, valuable, and useful book to Mr. Hutton, than whom a more competent man they could not have found."—*Iron.*

Templeton's Engineer's and Machinist's Assistant.

THE ENGINEER'S, MILLWRIGHT'S, and MACHINIST'S PRACTICAL ASSISTANT. A collection of Useful Tables, Rules and Data. By WILLIAM TEMPLETON. 7th Edition, with Additions. 18mo, 2s. 6d. cloth.

** OPINIONS OF THE PRESS.

"Occupies a foremost place among books of this kind. A more suitable present to an apprentice to any of the mechanical trades could not possibly be made."—*Building News.*

"A deservedly popular work. It should be in the 'drawer' of every mechanic."—*English Mechanic.*

Foley's Office Reference Book for Mechanical Engineers.

THE MECHANICAL ENGINEER'S REFERENCE BOOK, for Machine and Boiler Construction. In Two Parts. Part I. GENERAL ENGINEERING DATA. Part II. BOILER CONSTRUCTION. With 51 Plates and numerous Illustrations. By NELSON FOLEY, M.I.N.A. Second Edition, Revised throughout and much Enlarged. Folio, £3 3s. net half-bound. [*Just published*]

SUMMARY OF CONTENTS.

PART I.

MEASURES.—CIRCUMFERENCES AND AREAS, &c., SQUARES, CUBES, FOURTH POWERS.—SQUARE AND CUBE ROOTS.—SURFACE OF TUBES—RECIPROCALS.—LOGARITHMS. — MENSURATION. — SPECIFIC GRAVITIES AND WEIGHTS.—WORK AND POWER.—HEAT.—COMBUSTION.—EXPANSION AND CONTRACTION.—EXPANSION OF GASES.—STEAM.—STATIC FORCES.—GRAVITATION AND ATTRACTION.—MOTION AND COMPUTATION OF RESULTING FORCES.—ACCUMULATED WORK.—CENTRE AND RADIUS OF GYRATION.—MOMENT OF INERTIA.—CENTRE OF OSCILLATION.—ELECTRICITY.—STRENGTH OF MATERIALS.—ELASTICITY. — TEST SHEETS OF METALS.—FRICTION.—TRANSMISSION OF POWER.—FLOW OF LIQUIDS.—FLOW OF GASES.—AIR PUMPS, SURFACE CONDENSERS, &c.—SPEED OF STEAMSHIPS.—PROPELLERS. — CUTTING TOOLS.—FLANGES. — COPPER SHEETS AND TUBES.—SCREWS, NUTS, BOLT HEADS, &c.—VARIOUS RECIPES AND MISCELLANEOUS MATTER.

WITH DIAGRAMS FOR VALVE-GEAR, BELTING AND ROPES, DISCHARGE AND SUCTION PIPES, SCREW PROPELLERS, AND COPPER PIPES.

PART II.

TREATING OF, POWER OF BOILERS.—USEFUL RATIOS.—NOTES ON CONSTRUCTION. — CYLINDRICAL BOILER SHELLS. — CIRCULAR FURNACES.— FLAT PLATES.— STAYS.— GIRDERS.—SCREWS. — HYDRAULIC TESTS. — RIVETING.—BOILER SETTING, CHIMNEYS, AND MOUNTINGS.—FUELS, &c.—EXAMPLES OF BOILERS AND SPEEDS OF STEAMSHIPS.—NOMINAL AND NORMAL HORSE POWER.

WITH DIAGRAMS FOR ALL BOILER CALCULATIONS AND DRAWINGS OF MANY VARIETIES OF BOILERS.

*** OPINIONS OF THE PRESS.

"The book is one which every mechanical engineer may, with advantage to himself add to his library."—*Industries.*

"Mr. Foley is well fitted to compile such a work. . . . The diagrams are a great feature of the work. . . . Regarding the whole work, it may be very fairly stated that Mr. Foley has produced a volume which will undoubtedly fulfil the desire of the author and become indispensable to all mechanical engineers."—*Marine Engineer.*

"We have carefully examined this work, and pronounce it a most excellent reference book for the use of marine engineers."—*Journal of American Society of Naval Engineers.*

"A veritable monument of industry on the part of Mr. Foley, who has succeeded in producing what is simply invaluable to the engineering profession."—*Steamship.*

Coal and Speed Tables.

A POCKET BOOK OF COAL AND SPEED TABLES, for *Engineers and Steam-users.* By NELSON FOLEY, Author of "The Mechanical Engineer's Reference Book." Pocket-size, 3s. 6d. cloth.

"These tables are designed to meet the requirements of every-day use; they are of sufficient scope for most practical purposes, and may be commended to engineers and users of steam."—*Iron.*

"This pocket-book well merits the attention of the practical engineer. Mr. Foley has compiled a very useful set of tables, the information contained in which is frequently required by engineers, coal consumers and users of steam."—*Iron and Coal Trades Review.*

Steam Engine.

TEXT-BOOK ON THE STEAM ENGINE. With a Supplement on Gas Engines, and PART II. ON HEAT ENGINES. By T. M. GOODEVE, M.A., Barrister-at-Law, Professor of Mechanics at the Royal College of Science, London; Author of "The Principles of Mechanics," "The Elements of Mechanism," &c. Twelfth Edition, Enlarged. With numerous Illustrations. Crown 8vo, 6s. cloth.

"Professor Goodeve has given us a treatise on the steam engine which will bear comparison with anything written by Huxley or Maxwell, and we can award it no higher praise."—*Engineer.*

"Mr. Goodeve's text-book is a work of which every young engineer should possess himself."—*Mining Journal.*

Gas Engines.

ON GAS-ENGINES. With Appendix describing a Recent Engine with Tube Igniter. By T. M. GOODEVE, M.A. Crown 8vo, 2s. 6d. cloth.   [*Just published.*

"Like all Mr. Goodeve's writings, the present is no exception in point of general excellence It is a valuable little volume."—*Mechanical World.*

Steam Engine Design.

A HANDBOOK ON THE STEAM ENGINE, with especial Reference to Small and Medium-sized Engines. For the Use of Engine-Makers, Mechanical Draughtsmen, Engineering Students and Users of Steam Power. By HERMAN HAEDER, C.E. English Edition, Re-edited by the Author from the Second German Edition, and Translated, with considerable Additions and Alterations, by H. H. P. POWLES, A.M.I.C.E., M.I.M.E. With nearly 1,100 Illustrations. Crown 8vo, 9s. cloth.

"A perfect encyclopædia of the steam engine and its details, and one which must take a permanent place in English drawing-offices and workshops."—*A Foreman Pattern-maker.*

"This is an excellent book, and should be in the hands of all who are interested in the construction and design of medium sized stationary engines. . . . A careful study of its contents and the arrangement of the sections leads to the conclusion that there is probably no other book like it in this country. The volume aims at showing the results of practical experience, and it certainly may claim a complete achievement of this idea."—*Nature.*

"There can be no question as to its value. We cordially commend it to all concerned in the design and construction of the steam engine."—*Mechanical World.*

Steam Boilers.

A TREATISE ON STEAM BOILERS: Their Strength, Construction, and Economical Working. By ROBERT WILSON, C.E. Fifth Edition. 12mo, 6s. cloth.

"The best treatise that has ever been published on steam boilers."—*Engineer.*

"The author shows himself perfect master of his subject, and we heartily recommend all employing steam power to possess themselves of the work."—*Ryland's Iron Trade Circular.*

Boiler Chimneys.

BOILER AND FACTORY CHIMNEYS: Their Draught-Power and Stability. With a Chapter on *Lightning Conductors.* By ROBERT WILSON, A.I.C.E., Author of "A Treatise on Steam Boilers," &c. Second Edition. Crown 8vo, 3s. 6d. cloth.

"A valuable contribution to the literature of scientific building."—*The Builder.*

Boiler Making.

THE BOILER-MAKER'S READY RECKONER & ASSISTANT. With Examples of Practical Geometry and Templating, for the Use of Platers, Smiths and Riveters. By JOHN COURTNEY, Edited by D. K. CLARK, M.I.C.E. Third Edition, 480 pp., with 140 Illusts. Fcap. 8vo, 7s. half-bound.

"No workman or apprentice should be without this book."—*Iron Trade Circular.*

Locomotive Engine Development.

THE LOCOMOTIVE ENGINE AND ITS DEVELOPMENT. A Popular Treatise on the Gradual Improvements made in Railway Engines between 1803 and 1894. By CLEMENT E. STRETTON, C.E., Author of " Safe Railway Working," &c. Third Edition, Revised and Enlarged. With 95 Illustrations. Crown 8vo, 2s. 6d. cloth gilt.   [*Just published.*

"Students of railway history and all who are interested in the evolution of the modern locomotive will find much to attract and entertain in this volume."—*The Times.*

"The author of this work is well known to the railway world, and no one probably has a better knowledge of the history and development of the locomotive. The volume before us should be of value to all connected with the railway system of this country."—*Nature.*

Estimating for Engineering Work, &c.

ENGINEERING ESTIMATES, COSTS AND ACCOUNTS: A Guide to Commercial Engineering. With numerous Examples of Estimates and Costs of Millwright Work, Miscellaneous Productions, Steam Engines and Steam Boilers; and a Section on the Preparation of Costs Accounts. By A GENERAL MANAGER. Demy 8vo, 12s. cloth.

"This is an excellent and very useful book, covering subject matter in constant requisition to every factory and workshop. . . . The book is invaluable, not only to the young engineer, but also to the estimate department of every works."—*Builder.*

"We accord the work unqualified praise. The information is given in a plain, straightforward manner, and bears throughout evidence of the intimate practical acquaintance of the author with every phase of commercial engineering."—*Mechanical World.*

Fire Engineering.

FIRES, FIRE-ENGINES, AND FIRE-BRIGADES. With
a History of Fire-Engines, their Construction, Use, and Management: Remarks on Fire-Proof Buildings, and the Preservation of Life from Fire;
Statistics of the Fire Appliances in English Towns; Foreign Fire Systems;
Hints on Fire Brigades, &c. &c. By CHARLES F. T. YOUNG, C.E. With
numerous Illustrations. 544 pp., demy 8vo, £1 4s. cloth.

"To those interested in the subject of fires and fire apparatus, we most heartily commend this
book. It is the only English work we now have upon the subject."—*Engineering.*

"It displays much evidence of careful research; and Mr. Young has put his facts neatly
together. His acquaintance with the practical details of the construction of steam fire engines,
old and new, and the conditions with which it is necessary they should comply, is accurate and
full."—*Engineer.*

Boilermaking.

PLATING AND BOILERMAKING: A Practical Handbook
for Workshop Operations. By JOSEPH G. HORNER, A.M.I.M.E. (Foreman
Pattern-Maker), Author of "Pattern Making," &c. 380 pages, with 338
Illustrations. Crown 8vo, 7s. 6d. cloth.   [*Just published.*

"A thoroughly practical, plainly-written treatise. The volume merits commendation. The
author's long experience enables him to write with full knowledge of his subject."—*Glasgow Herald.*

Engineering Construction.

PATTERN-MAKING: A Practical Treatise, embracing the Main
Types of Engineering Construction, and including Gearing, both Hand and
Machine made, Engine Work, Sheaves and Pulleys, Pipes and Columns,
Screws, Machine Parts, Pumps and Cocks, the Moulding of Patterns in
Loam and Greensand, &c., together with the methods of Estimating the
weight of Castings; to which is added an Appendix of Tables for Workshop
Reference. By JOSEPH G. HORNER, A.M.I.M.E. (Foreman Pattern-Maker).
Second Edition, thoroughly Revised and much Enlarged. With upwards of
450 Illustrations. Crown 8vo, 7s. 6d. cloth.   [*Just published.*

"A well-written technical guide, evidently written by a man who understands and has practised what he has written about. . . . We cordially recommend it to engineering students, young
journeymen, and others desirous of being initiated into the mysteries of pattern-making."—*Builder.*

"More than 450 illustrations help to explain the text, which is, however, always clear and explicit, thus rendering the work an excellent *vade mecum* for the apprentice who desires to become
master of his trade."—*English Mechanic.*

Dictionary of Mechanical Engineering Terms.

*LOCKWOOD'S DICTIONARY OF TERMS USED IN THE
PRACTICE OF MECHANICAL ENGINEERING,* embracing those current
in the Drawing Office, Pattern Shop, Foundry, Fitting, Turning, Smith's and
Boiler Shops, &c. &c. Comprising upwards of 6,000 Definitions. Edited by
JOSEPH G. HORNER, A.M.I.M.E. (Foreman Pattern-Maker), Author of "Pat-
tern Making." Second Edition, Revised. Crown 8vo, 7s. 6d. cloth.

"Just the sort of handy dictionary required by the various trades engaged in mechanical engineering. The practical engineering pupil will find the book of great value in his studies, and
every foreman engineer and mechanic should have a copy."—*Building News.*

"Not merely a dictionary, but, to a certain extent, also a most valuable guide. It strikes us as
a happy idea to combine with a definition of the phrase useful information on the subject of which
it treats."—*Machinery Market.*

Mill Gearing.

TOOTHED GEARING: A Practical Handbook for Offices and
Workshops. By JOSEPH G. HORNER, A.M.I.M.E. (Foreman Pattern-Maker),
Author of "Pattern Making," &c. With 184 Illustrations. Crown 8vo, 6s.
cloth.   [*Just published.*

SUMMARY OF CONTENTS.

CHAP. I. PRINCIPLES.—II. FORMA-
TION OF TOOTH PROFILES.—III. PRO-
PORTIONS OF TEETH.—IV. METHODS
OF MAKING TOOTH FORMS.—V. INVO-
LUTE TEETH. — VI. SOME SPECIAL
TOOTH FORMS.—VII. BEVEL WHEELS.
— VIII. SCREW GEARS. — IX. WORM
GEARS.—X. HELICAL WHEELS.—XI.
SKEW BEVELS.—XII. VARIABLE AND
OTHER GEARS.—XIII. DIAMETRICAL
PITCH.—XIV. THE ODONTOGRAPH.—
XV. PATTERN GEARS.—XVI. MACHINE
MOULDING GEARS.—XVII. MACHINE
CUT GEARS.—XVIII. PROPORTION OF
WHEELS.

"We must give the book our unqualified praise for its thoroughness of treatment, and we can
heartily recommend it to all interested as the most practical book on the subject yet written."—
Mechanical World.

Stone-working Machinery.

STONE-WORKING MACHINERY, and the Rapid and Economical Conversion of Stone. With Hints on the Arrangement and Management of Stone Works. By M. POWIS BALE, M.I.M.E. With Illusts. Crown 8vo, 9s.

"The book should be in the hands of every mason or student of stone-work."—*Colliery Guardian*.

"A capital handbook for all who manipulate stone for building or ornamental purposes."—*Machinery Market*.

Pump Construction and Management.

PUMPS AND PUMPING : A Handbook for Pump Users. Being Notes on Selection, Construction and Management. By M. POWIS BALE, M.I.M.E., Author of "Woodworking Machinery," "Saw Mills," &c. Second Edition, Revised. Crown 8vo, 2s. 6d. cloth.

"The matter is set forth as concisely as possible. In fact, condensation rather than diffuseness has been the author's aim throughout; yet he does not seem to have omitted anything likely to be of use."—*Journal of Gas Lighting*.

"Thoroughly practical and simply and clearly written."—*Glasgow Herald*.

Milling Machinery, etc.

MILLING MACHINES AND PROCESSES : A Practical Treatise on Shaping Metals by Rotary Cutters, including Information on Making and Grinding the Cutters. By PAUL N. HASLUCK, Author of "Lathe-work," "Handybooks for Handicrafts," &c. With upwards of 300 Engravings, including numerous Drawings by the Author. Large crown 8vo, 352 pages, 12s. 6d. cloth.

"A new departure in engineering literature. . . We can recommend this work to all interested in milling machines; it is what it professes to be—a practical treatise."—*Engineer*.

"A capital and reliable book, which will no doubt be of considerable service, both to those who are already acquainted with the process as well as to those who contemplate its adoption."
Industries.

Turning.

LATHE-WORK : A Practical Treatise on the Tools, Appliances, and Processes employed in the Art of Turning. By PAUL N. HASLUCK. Fifth Edition, Revised and Enlarged Cr. 8vo, 5s. cloth.

"Written by a man who knows, not only how work ought to be done, but who also knows how to do it, and how to convey his knowledge to others. To all turners this book would be valuable."—*Engineering*.

"We can safely recommend the work to young engineers. To the amateur it will simply be invaluable. To the student it will convey a great deal of useful information."—*Engineer*.

Screw-Cutting.

SCREW THREADS : And Methods of Producing Them. With Numerous Tables, and complete directions for using Screw-Cutting Lathes. By PAUL N. HASLUCK, Author of "Lathe-Work," &c. With Seventy-four Illustrations. Third Edition, Revised and Enlarged. Waistcoat-pocket size, 1s. 6d. cloth.

"Full of useful information, hints and practical criticism. Taps, dies and screwing-tools generally are illustrated and their action described."—*Mechanical World*.

"It is a complete compendium of all the details of the screw cutting lathe; in fact a *multum in parvo* on all the subjects it treats upon."—*Carpenter and Builder*.

Smith's Tables for Mechanics, etc.

TABLES, MEMORANDA, AND CALCULATED RESULTS, FOR MECHANICS, ENGINEERS, ARCHITECTS, BUILDERS, etc. Selected and Arranged by FRANCIS SMITH. Fifth Edition, thoroughly Revised and Enlarged, with a New Section of ELECTRICAL TABLES, FORMULÆ, and MEMORANDA. Waistcoat-pocket size, 1s. 6d. limp leather.

"It would, perhaps, be as difficult to make a small pocket-book selection of notes and formulæ to suit ALL engineers as it would be to make a universal medicine; but Mr. Smith's waistcoat-pocket collection may be looked upon as a successful attempt."—*Engineer*.

"The best example we have ever seen of 270 pages of useful matter packed into the dimensions of a card-case."—*Building News*. "A veritable pocket treasury of knowledge."—*Iron*.

French-English Glossary for Engineers, etc.

A POCKET GLOSSARY of TECHNICAL TERMS : ENGLISH-FRENCH, FRENCH-ENGLISH ; with Tables suitable for the Architectural, Engineering, Manufacturing and Nautical Professions. By JOHN JAMES FLETCHER, Engineer and Surveyor. Second Edition, Revised and Enlarged, 200 pp. Waistcoat-pocket size, 1s. 6d. limp leather.

"It is a very great advantage for readers and correspondents in France and England to have so large a number of the words relating to engineering and manufactures collected in a lilliputian volume. The little book will be useful both to students and travellers."—*Architect*.

"The glossary of terms is very complete, and many of the tables are new and well arranged. We cordially commend the book."—*Mechanical World*.

Year-Book of Engineering Formulæ, &c.

THE ENGINEER'S YEAR-BOOK FOR 1895. Comprising Formulæ, Rules, Tables, Data and Memoranda in Civil, Mechanical, Electrical, Marine and Mine Engineering. By H. R. KEMPE, A.M. Inst.C.E., M.I.E.E., Technical Officer of the Engineer-in-Chief's Office, General Post Office, London, Author of "A Handbook of Electrical Testing," "The Electrical Engineer's Pocket-Book," &c. With 750 Illustrations, specially Engraved for the work. Crown 8vo, 650 pages, 8s. leather. [*Just published.*

"Represents an enormous quantity of work, and forms a desirable book of reference."—*The Engineer.*

"The book is distinctly in advance of most similar publications in this country."—*Engineering*

"This valuable and well-designed book of reference meets the demands of all descriptions of engineers."—*Saturday Review.*

"Teems with up-to-date information in every branch of engineering and construction."—*Building News.*

"The needs of the engineering profession could hardly be supplied in a more admirable, complete and convenient form. To say that it more than sustains all comparisons is praise of the highest sort, and that may justly be said of it.'—*Mining Journal.*

"There is certainly room for the new comer, which supplies explanations and directions, as well as formulæ and tables. It deserves to become one of the most successful of the technical annuals."—*Architect.*

"Brings together with great skill all the technical information which an engineer has to use day by day. It is in every way admirably equipped, and is sure to prove successful."—*Scotsman.*

"The up-to-dateness of Mr. Kempe's compilation is a quality that will not be lost on the busy people for whom the work is intended."—*Glasgow Herald.*

Portable Engines.

THE PORTABLE ENGINE; ITS CONSTRUCTION AND MANAGEMENT. A Practical Manual for Owners and Users of Steam Engines generally. By WILLIAM DYSON WANSBROUGH. With 90 Illustrations. Crown 8vo, 3s. 6d. cloth.

"This is a work of value to those who use steam machinery. . . . Should be read by everyone who has a steam engine, on a farm or elsewhere."—*Mark Lane Express.*

"We cordially commend this work to buyers and owners of steam engines, and to those who have to do with their construction or use."—*Timber Trades Journal.*

"Such a general knowledge of the steam engine as Mr. Wansbrough furnishes to the reader should be acquired by all intelligent owners and others who use the steam engine."—*Building News.*

"An excellent text-book of this useful form of engine. 'The Hints to Purchasers' contain a good deal of commonsense and practical wisdom."—*English Mechanic.*

Iron and Steel.

"IRON AND STEEL": A Work for the Forge, Foundry, Factory, and Office. Containing ready, useful, and trustworthy Information for Ironmasters and their Stock-takers; Managers of Bar, Rail, Plate, and Sheet Rolling Mills; Iron and Metal Founders; Iron Ship and Bridge Builders; Mechanical, Mining, and Consulting Engineers; Architects, Contractors, Builders, and Professional Draughtsmen. By CHARLES HOARE, Author of "The Slide Rule," &c. Eighth Edition, Revised throughout and considerably Enlarged. 32mo, 6s. leather.

"For comprehensiveness the book has not its equal."—*Iron.*

"One of the best of the pocket books."—*English Mechanic.*

"We cordially recommend this book to those engaged in considering the details of all kinds of iron and steel works."—*Naval Science.*

Elementary Mechanics.

CONDENSED MECHANICS. A Selection of Formulæ, Rules, Tables, and Data for the Use of Engineering Students, Science Classes, &c. In Accordance with the Requirements of the Science and Art Department. By W. G. CRAWFORD HUGHES, A.M.I.C.E. Crown 8vo, 2s. 6d. cloth.

"The book is well fitted for those who are either confronted with practical problems in their work, or are preparing for examination and wish to refresh their knowledge by going through their formulæ again."—*Marine Engineer.*

"It is well arranged, and meets the wants of those for whom it is intended."—*Railway News.*

Steam.

THE SAFE USE OF STEAM. Containing Rules for Unprofessional Steam-users. By an ENGINEER. Sixth Edition. Sewed, 6d.

"If steam-users would but learn this little book by heart, boiler explosions would become sensations by their rarity."—*English Mechanic.*

Warming.

HEATING BY HOT WATER; with Information and Suggestions on the best Methods of Heating Public, Private and Horticultural Buildings. By WALTER JONES. Second Edition. With 96 Illustrations. Crown 8vo, 2s. 6d. net.

"We confidently recommend all interested in heating by hot water to secure a copy of this valuable little treatise."—*The Plumber and Decorator.*

CIVIL ENGINEERING, SURVEYING, etc.

Water Supply and Water-Works.

THE WATER SUPPLY OF TOWNS AND THE CON-
STRUCTION OF WATER-WORKS: A Practical Treatise for the Use of
Engineers and Students of Engineering. By W. K. BURTON, A.M.Inst C E.,
Professor of Sanitary Engineering in the Imperial University, Tokyo, Japan,
and Consulting Engineer to the Tokyo Water-works. With an Appendix on
the Effects of Earthquakes on Waterworks, by JOHN MILNE, F.R.S., Pro-
fessor of Mining in the Imperial University of Japan. With numerous
Plates and Illustrations. Super-royal 8vo, 25s. buckram. [*Just published.*

"The whole art of waterworks construction is dealt with in a clear and comprehensive fashion
in this handsome volume. . . . Mr. Burton's practical treatise shows in all its sections the fruit
of independent study and individual experience. It is largely based upon his own practice in the
branch of engineering of which it treats, and with such a basis a treatise can scarcely fail to be suggestive and useful."—*Saturday Review.*

"Professor Burton's book is sure of a warm welcome among engineers. It is written in clear
and vigorous language and forms an exhaustive treatise on a branch of engineering the claims of
which it would be difficult to over-estimate."—*Scotsman.*

"The subjects seem to us to be ably discussed, with a practical aim to meet the requirements
of all its probable readers. The volume is well got up, and the illustrations are excellent."
The Lancet.

The Water Supply of Cities and Towns.

A COMPREHENSIVE TREATISE on the WATER-SUPPLY
OF CITIES AND TOWNS. By WILLIAM HUMBER, A-M.Inst.C.E., and
M. Inst. M.E., Author of "Cast and Wrought Iron Bridge Construction,"
&c. &c. Illustrated with 50 Double Plates, 1 Single Plate, Coloured
Frontispiece, and upwards of 250 Woodcuts, and containing 400 pages of
Text. Imp. 4to, £6 6s. elegantly and substantially half-bound in morocco.

List of Contents.

I. Historical Sketch of some of the means
that have been adopted for the Supply of Water
to Cities and Towns.—II. Water and the Foreign Matter usually associated with it.—III.
Rainfall and Evaporation.—IV. Springs and
the water-bearing formations of various districts.—V. Measurement and Estimation of the
flow of Water—VI. On the Selection of the
Source of Supply.—VII. Wells.—VIII. Reser-
voirs.—IX. The Purification of Water.—X.
Pumps. — XI. Pumping Machinery. — XII.
Conduits.—XIII. Distribution of Water.—XIV.
Meters, Service Pipes, and House Fittings.—
XV. The Law and Economy of Water Works.
XVI. Constant and Intermittent Supply.—
XVII. Description of Plates. — Appendices,
giving Tables of Rates of Supply, Velocities,
&c. &c., together with Specifications of several
Works illustrated, among which will be found :
Aberdeen, Bideford, Canterbury, Dundee,
Halifax, Lambeth, Rotherham, Dublin, and
others.

"The most systematic and valuable work upon water supply hitherto produced in English, or
in any other language. . . . Mr. Humber's work is characterised almost throughout by an
exhaustiveness much more distinctive of French and German than of English technical treatises."
—*Engineer.*

"We can congratulate Mr. Humber on having been able to give so large an amount of information on a subject so important as the water supply of cities and towns. The plates, fifty in
number, are mostly drawings of executed works, and alone would have commanded the attention
of every engineer whose practice may lie in this branch of the profession."—*Builder.*

Water Supply.

RURAL WATER SUPPLY: A Practical Handbook on the
Supply of Water and Construction of Waterworks for small Country Districts.
By ALLAN GREENWELL, A.M.I.C.E., and W. T. CURRY, A.M.I.C.E., F.G.S.
With Illustrations. Crown 8vo, 5s. cloth. [*Just ready.*

Hydraulic Tables.

HYDRAULIC TABLES, CO-EFFICIENTS, and FORMULÆ
for finding the Discharge of Water from Orifices, Notches, Weirs, Pipes, and
Rivers. With New Formulæ, Tables, and General Information on Rainfall,
Catchment-Basins, Drainage, Sewerage, Water Supply for Towns and Mill
Power. By JOHN NEVILLE, Civil Engineer, M.R.I.A. Third Ed., carefully
Revised, with considerable Additions. Numerous Illusts. Cr. 8vo, 14s. cloth.

"Alike valuable to students and engineers in practice ; its study will prevent the annoyance of
avoidable failures, and assist them to select the readiest means of successfully carrying out any
given work connected with hydraulic engineering."—*Mining Journal.*

"It is, of all English books on the subject, the one nearest to completeness. . . . From the
good arrangement of the matter, the clear explanations, and abundance of formulæ, the carefully
calculated tables, and, above all, the thorough acquaintance with both theory and construction,
which is displayed from first to last, the book will be found to be an acquisition."—*Architect.*

Hydraulics.

HYDRAULIC MANUAL. Consisting of Working Tables and Explanatory Text. Intended as a Guide in Hydraulic Calculations and Field Operations. By LEWIS D'A. JACKSON, Author of "Aid to Survey Practice," "Modern Metrology," &c. Fourth Edition, Enlarged. Large cr. 8vo, 16s. cl.

"The author has had a wide experience in hydraulic engineering and has been a careful observer of the facts which have come under his notice, and from the great mass of material at his command he has constructed a manual which may be accepted as a trustworthy guide to this branch of the engineer's profession. We can heartily recommend this volume to all who desire to be acquainted with the latest development of this important subject."—*Engineering.*

"The standard-work in this department of mechanics."—*Scotsman.*

"The most useful feature of this work is its freedom from what is superannuated, and its thorough adoption of recent experiments; the text is, in fact, in great part a short account of the great modern experiments."—*Nature.*

Water Storage, Conveyance, and Utilisation.

WATER ENGINEERING : A Practical Treatise on the Measurement, Storage, Conveyance, and Utilisation of Water for the Supply of Towns, for Mill Power, and for other Purposes. By CHARLES SLAGG, A.M.Inst.C.E., Author of "Sanitary Work in the Smaller Towns, and in Villages," &c. Second Edition. With numerous Illustrations. Crown 8vo, 7s. 6d. cloth.

"As a small practical treatise on the water supply of towns, and on some applications of water-power, the work is in many respects excellent."—*Engineering.*

"The author has collated the results deduced from the experiments of the most eminent authorities, and has presented them in a compact and practical form, accompanied by very clear and detailed explanations. . . . The application of water as a motive power is treated very carefully and exhaustively."—*Builder.*

"For anyone who desires to begin the study of hydraulics with a consideration of the practical applications of the science there is no better guide."—*Architect.*

Drainage.

ON THE DRAINAGE OF LANDS, TOWNS, AND BUILDINGS. By G. D. DEMPSEY, C.E., Author of "The Practical Railway Engineer," &c. Revised, with large Additions on RECENT PRACTICE IN DRAINAGE ENGINEERING, by D. KINNEAR CLARK, M.Inst.C.E. Author of "Tramways: Their Construction and Working," "A Manual of Rules, Tables, and Data for Mechanical Engineers," &c. Second Edition, Corrected. Fcap. 8vo, 5s. cloth.

"The new matter added to Mr. Dempsey's excellent work is characterised by the comprehensive grasp and accuracy of detail for which the name of Mr. D. K. Clark is a sufficient voucher."—*Athenæum.*

"As a work on recent practice in drainage engineering, the book is to be commended to ll who are making that branch of engineering science their special study."—*Iron.*

"A comprehensive manual on drainage engineering, and a useful introduction to the student."—*Building News.*

River Engineering.

RIVER BARS: *The Causes of their Formation, and their Treatment by "Induced Tidal Scour;"* with a Description of the Successful Reduction by this Method of the Bar at Dublin. By I. J. MANN, Assist. Eng. to the Dublin Port and Docks Board. Royal 8vo, 7s. 6d. cloth.

"We recommend all interested in harbour works—and, indeed, those concerned in the improvements of rivers generally—to read Mr. Mann's interesting work on the treatment of river bars."—*Engineer.*

Tramways and their Working.

TRAMWAYS : THEIR CONSTRUCTION AND WORKING. Embracing a Comprehensive History of the System ; with an exhaustive Analysis of the various Modes of Traction, including Horse-Power, Steam, Cable Traction, Electric Traction, &c.; a Description of the Varieties of Rolling Stock; and ample Details of Cost and Working Expenses. New Edition, Thoroughly Revised, and Including the Progress recently made in Tramway Construction, &c. &c. By D. KINNEAR CLARK. M.Inst.C.E. With numerous Illustrations and Folding Plates. In One Volume, 8vo, 780 pages, price 28s., bound in buckram.   [*Just published.*

"All interested in tramways must refer to it, as all railway engineers have turned to the author's work 'Railway Machinery.'"—*Engineer.*

"An exhaustive and practical work on tramways, in which the history of this kind of locomotion, and a description and cost of the various modes of laying tramways, are to be found."—*Building News.*

"The best form of rails, the best mode of construction, and the best mechanical appliances are so fairly indicated in the work under review, that any engineer about to construct a tramway will be enabled at once to obtain the practical information which will be of most service to him."—*Athenæum.*

Student's Text-Book on Surveying.

PRACTICAL SURVEYING : A Text-Book for Students preparing for Examination or for Survey-work in the Colonies. By GEORGE W. USILL, A.M.I.C.E., Author of "The Statistics of the Water Supply of Great Britain." With Four Lithographic Plates and upwards of 330 Illustrations. Third Edition, Revised and Enlarged. Including Tables of Natural Sines, Tangents, Secants, &c. Crown 8vo, 7s. 6d. cloth ; or, on THIN PAPER, bound in limp leather, gilt edges, rounded corners, for pocket use, 12s. 6d.

" The best forms of instruments are described as to their construction, uses and modes of employment, and there are innumerable hints on work and equipment such as the author, in his experience as surveyor, draughtsman, and teacher, has found necessary, and which the student in his inexperience will find most serviceable."—*Engineer.*

" The latest treatise in the English language on surveying, and we have no hesitation in saying that the student will find it a better guide than any of its predecessors . . . . . Deserves to be recognised as the first book which should be put in the hands of a pupil of Civil Engineering, and every gentleman of education who sets out for the Colonies would find it well to have a copy."—*Architect.*

Survey Practice.

AID TO SURVEY PRACTICE, for Reference in Surveying, Levelling, and Setting-out ; and in Route Surveys of Travellers by Land and Sea. With Tables, Illustrations, and Records. By LEWIS D'A. JACKSON, A.M.I.C.E., Author of "Hydraulic Manual," "Modern Metrology," &c. Second Edition, Enlarged. Large crown 8vo, 12s. 6d. cloth.

" A valuable *vade-mecum* for the surveyor. We can recommend this book as containing an admirable supplement to the teaching of the accomplished surveyor."— *Athenæum.*

" As a text-book we should advise all surveyors to place it in their libraries, and study well the matured instructions afforded in its pages."—*Colliery Guardian.*

" The author brings to his work a fortunate union of theory and practical experience which, aided by a clear and lucid style of writing, renders the book a very useful one."—*Builder.*

Surveying, Land and Marine.

LAND AND MARINE SURVEYING, in Reference to the Preparation of Plans for Roads and Railways ; Canals, Rivers, Towns' Water Supplies ; Docks and Harbours. With Description and Use of Surveying Instruments. By W. D. HASKOLL, C.E., Author of "Bridge and Viaduct Construction," &c. Second Edition, Revised, with Additions. Large cr. 8vo, 9s. cl.

" This book must prove of great value to the student. We have no hesitation in recommending it, feeling assured that it will more than repay a careful study."—*Mechanical World.*

" A most useful and well arranged book. We can strongly recommend it as a carefully-written and valuable text-book. It enjoys a well-deserved repute among surveyors."—*Builder.*

" This volume cannot fail to prove of the utmost practical utility. It may be safely recommended to all students who aspire to become clean and expert surveyors."—*Mining Journal.*

Field-Book for Engineers.

THE ENGINEER'S, MINING SURVEYOR'S, AND CONTRACTOR'S FIELD-BOOK. Consisting of a Series of Tables, with Rules, Explanations of Systems, and use of Theodolite for Traverse Surveying and Plotting the Work with minute accuracy by means of Straight Edge and Set Square only ; Levelling with the Theodolite, Casting-out and Reducing Levels to Datum, and Plotting Sections in the ordinary manner ; setting-out Curves with the Theodolite by Tangential Angles and Multiples, with Right and Left-hand Readings of the Instrument ; Setting-out Curves without Theodolite, on the System of Tangential Angles by sets of Tangents and Offsets ; and Earthwork Tables to 80 feet deep, calculated for every 6 inches in depth. By W. D. HASKOLL, C.E. Fourth Edition. Crown 8vo, 12s. cloth.

" The book is very handy ; the separate tables of sines and tangents to every minute will make it useful for many other purposes, the genuine traverse tables existing all the same."—*Athenæum.*

" Every person engaged in engineering field operations will estimate the importance of such a work and the amount of valuable time which will be saved by reference to a set of reliable tables prepared with the accuracy and fulness of those given in this volume."—*Railway News.*

Levelling.

A TREATISE ON THE PRINCIPLES AND PRACTICE OF LEVELLING. Showing its Application to purposes of Railway and Civil Engineering, in the Construction of Roads ; with Mr. TELFORD'S Rules for the same. By FREDERICK W. SIMMS, F.G.S., M.Inst.C.E. Seventh Edition, with the addition of LAW'S Practical Examples for Setting-out Railway Curves, and TRAUTWINE'S Field Practice of Laying-out Circular Curves. With 7 Plates and numerous Woodcuts. 8vo, 8s. 6d. cloth. **** TRAUTWINE on Curves may be had separate, 5s.

" The text-book on levelling in most of our engineering schools and colleges. . . . The publishers have rendered a substantial service to the profession, especially to the younger members, by bringing out the present edition of Mr. Simms's useful book."—*Engineer.*

Trigonometrical Surveying.

AN OUTLINE OF THE METHOD OF CONDUCTING A TRIGONOMETRICAL SURVEY, for the Formation of Geographical and Topographical Maps and Plans, Military Reconnaissance, Levelling, &c., with Useful Problems, Formulæ, and Tables. By Lieut.-General FROME, R.E. Fourth Edition, Revised and partly Re-written by Major General Sir CHARLES WARREN, G.C.M.G., R.E. With 19 Plates and 115 Woodcuts. Royal 8vo, 16s. cloth.

"The simple fact that a fourth edition has been called for is the best testimony to its merits. No words of praise from us can strengthen the position so well and so steadily maintained by this work. Sir Charles Warren has revised the entire work, and made such additions as were necessary to bring every portion of the contents up to the present date."—*Broad Arrow.*

Curves, Tables for Setting-out.

TABLES OF TANGENTIAL ANGLES AND MULTIPLES for Setting-out Curves from 5 to 200 Radius. By ALEXANDER BEAZELEY, M.Inst.C.E. Fourth Edition. Printed on 48 Cards, and sold in a cloth box, waistcoat-pocket size, 3s. 6d.

"Each table is printed on a small card, which, being placed on the theodolite, leaves the hands free to manipulate the instrument—no small advantage as regards the rapidity of work."—*Engineer.*
"Very handy; a man may know that all his day's work must fall on two of these cards, which he puts into his own card-case, and leaves the rest behind."—*Athenæum.*

Earthwork.

EARTHWORK TABLES. Showing the Contents in Cubic Yards of Embankments, Cuttings, &c., of Heights or Depths up to an average of 80 feet. By JOSEPH BROADBENT, C.E., and FRANCIS CAMPIN, C.E. Crown 8vo, 5s. cloth.

"The way in which accuracy is attained, by a simple division of each cross section into three elements, two in which are constant and one variable, is ingenious."—*Athenæum.*

Earthwork, Measurement of.

A MANUAL ON EARTHWORK. By ALEX. J. S. GRAHAM, C.E. With numerous Diagrams. Second Edition. 18mo, 2s. 6d. cloth.

"A great amount of practical information, very admirably arranged, and available for rough estimates, as well as for the more exact calculations required in the engineer's and contractor's offices."—*Artisan.*

Tunnelling.

PRACTICAL TUNNELLING. Explaining in detail the Setting-out of the works, Shaft-sinking and Heading-driving, Ranging the Lines and Levelling underground, Sub-Excavating, Timbering, and the Construction of the Brickwork of Tunnels, with the amount of Labour required for, and the Cost of, the various portions of the work. By FREDERICK W. SIMMS, F.G.S., M.Inst.C.E. Third Edition, Revised and Extended by D. KINNEAR CLARK, M.Inst.C.E. Imperial 8vo, with 21 Folding Plates and numerous Wood Engravings, 30s. cloth.

"The estimation in which Mr. Simms's book on tunnelling has been held for over thirty years cannot be more truly expressed than in the words of the late Prof. Rankine:—'The best source of information or the subject of tunnels is Mr. F. W. Simms's work on Practical Tunnelling.'"—*Architect.*
"It has been regarded from the first as a text-book of the subject. . . . Mr. Clark has added immensely to the value of the book."—*Engineer.*

Tunnel Shafts.

THE CONSTRUCTION OF LARGE TUNNEL SHAFTS: A Practical and Theoretical Essay. By J. H. WATSON BUCK, M.Inst.C.E., Resident Engineer, London and North-Western Railway. Illustrated with Folding Plates. Royal 8vo, 12s. cloth.

"Many of the methods given are of extreme practical value to the mason; and the observations on the form of arch, the rules for ordering the stone, and the construction of the templates will be found of considerable use. We commend the book to the engineering profession."—*Building News.*
"Will be regarded by civil engineers as of the utmost value, and calculated to save much time and obviate many mistakes."—*Colliery Guardian.*

Oblique Bridges.

A PRACTICAL AND THEORETICAL ESSAY ON OBLIQUE BRIDGES. With 13 large Plates. By the late GEORGE WATSON BUCK, M.I.C.E. Fourth Edition, revised by his Son, J. H. WATSON BUCK, M.I.C.E.; and with the addition of Description to Diagrams for Facilitating the Construction of Oblique Bridges, by W. H. BARLOW, M.I.C.E. Roy. 8vo, 12s cl.

"The standard text-book for all engineers regarding skew arches is Mr. Buck's treatise, and it would be impossible to consult a better."—*Engineer.*
"Mr. Buck's treatise is recognised as a standard text-book, and his treatment has divested the subject of many of the intricacies supposed to belong to it. As a guide to the engineer and architect, on a confessedly difficult subject, Mr. Buck's work is unsurpassed."—*Building News.*

Cast and Wrought Iron Bridge Construction.

A COMPLETE AND PRACTICAL TREATISE ON CAST AND WROUGHT IRCN BRIDGE CONSTRUCTION, including Iron Foundations. In Three Parts—Theoretical, Practical, and Descriptive. By WILLIAM HUMBER, A.M.Inst.C.E., and M.Inst.M.E. Third Edition, Revised and much improved, with 115 Double Plates (20 of which now first appear in this edition), and numerous Additions to the Text. In Two Vols., imp. 4to, £6 16s. 6d. half-bound in morocco.

"A very valuable contribution to the standard literature of civil engineering. In addition to elevations, plans and sections, large scale details are given which very much enhance the instructive worth of those illustrations."—*Civil Engineer and Architect's Journal.*

"Mr. Humber's state'y volumes, lately issued—in which the most important bridges erected during the last five years, under the direction of the late Mr. Brunel, Sir W. Cubitt, Mr. Hawkshaw, Mr. Page, Mr. Fowler, Mr. Hemans, and others among our most eminent engineers, are drawn and specified in great detail."—*Engineer.*

Oblique Arches.

A PRACTICAL TREATISE ON THE CONSTRUCTION OF OBLIQUE ARCHES. By JOHN HART. Third Edition, with Plates. Imperial 8vo, 8s. cloth.

Statics, Graphic and Analytic.

GRAPHIC AND ANALYTIC STATICS, in their Practical Application to the Treatment of Stresses in Roofs, Solid Girders, Lattice, Bowstring and Suspension Bridges, Braced Iron Arches and Piers, and other Frameworks. By R. HUDSON GRAHAM, C.E. Containing Diagrams and Plates to Scale. With numerous Examples, many taken from existing Structures. Specially arranged for Class-work in Colleges and Universities. Second Edition, Revised and Enlarged. 8vo. 16s. cloth.

"Mr. Graham's book will find a place wherever graphic and analytic statics are used or studied."
—*Engineer.*

"The work is excellent from a practical point of view, and has evidently been prepared with much care. The directions for working are ample, and are illustrated by an abundance of well-selected examples. It is an excellent text-book for the practical draughtsman."—*Athenæum.*

Girders, Strength of.

GRAPHIC TABLE FOR FACILITATING THE COMPUTATION OF THE WEIGHTS OF WROUGHT IRON AND STEEL GIRDERS, etc., for Parliamentary and other Estimates. By J. H. WATSON BUCK, M.Inst.C.E. On a Sheet, 2s. 6d.

Strains, Calculation of.

A HANDY BOOK FOR THE CALCULATION OF STRAINS IN GIRDERS AND SIMILAR STRUCTURES, AND THEIR STRENGTH. Consisting of Formulæ and Corresponding Diagrams, with numerous details for Practical Application, &c. By WILLIAM HUMBER, A-M.Inst.C.E., &c. Fifth Edition. Crown 8vo, nearly 100 Woodcuts and 3 Plates, 7s. 6d. cloth.

"The formulæ are neatly expressed, and the diagrams good."—*Athenæum.*

"We heartily commend this really handy book to our engineer and architect readers."—*English Mechanic.*

Trusses.

TRUSSES OF WOOD AND IRON. Practical Applications of Science in Determining the Stresses, Breaking Weights, Safe Loads, Scantlings, and Details of Construction, with Complete Working Drawings. By WILLIAM GRIFFITHS, Surveyor, Assistant Master, Tranmere School of Science and Art. Oblong 8vo, 4s. 6d. cloth.

"This handy little book enters so minutely into every detail connected with the construction of roof trusses, that no student need be ignorant of these matters."—*Practical Engineer.*

Strains in Ironwork.

THE STRAINS ON STRUCTURES OF IRONWORK; with Practical Remarks on Iron Construction. By F. W. SHEILDS, M.Inst.C.E. Second Edition, with 5 Plates. Royal 8vo, 5s. cloth.

"The student cannot find a better little book on this subject."—*Engineer.*

Barlow's Strength of Materials, enlarged by Humber.

A TREATISE ON THE STRENGTH OF MATERIALS; with Rules for Application in Architecture, the Construction of Suspension Bridges, Railways, &c. By PETER BARLOW, F.R.S. A New Edition, Revised by his Sons, P. W. BARLOW, F.R.S., and W. H. BARLOW, F.R.S.; to which are added, Experiments by HODGKINSON, FAIRBAIRN, and KIRKALDY; and Formulæ for Calculating Girders, &c. Arranged and Edited by WM. HUMBER, A-M. Inst.C.E. Demy 8vo, 400 pp., with 19 large Plates and numerous Woodcuts, 18s. cloth.

"Valuable alike to the student, tyro, and the experienced practitioner, it will always rank in future, as it has hitherto done, as the standard treatise on that particular subject."—*Engineer.*
"There is no greater authority than Barlow."—*Building News.*
"As a scientific work of the first class, it deserves a foremost place on the bookshelves of every civil engineer and practical mechanic."—*English Mechanic.*

Cast Iron and other Metals, Strength of.

A PRACTICAL ESSAY ON THE STRENGTH OF CAST IRON AND OTHER METALS. By THOMAS TREDGOLD, C.E. Fifth Edition, including HODGKINSON'S Experimental Researches. 8vo, 12s. cloth.

Practical Mathematics.

MATHEMATICS FOR PRACTICAL MEN: Being a Commonplace Book of Pure and Mixed Mathematics. Designed chiefly for the use of Civil Engineers, Architects and Surveyors. By OLINTHUS GREGORY, LL.D., F.R.A.S., Enlarged by HENRY LAW, C.E. 4th Edition, carefully Revised by J. R. YOUNG, formerly Professor of Mathematics, Belfast College. With 13 Plates. 8vo, £1 1s. cloth.

"The engineer or architect will here find ready to his hand rules for solving nearly every mathematical difficulty that may arise in his practice. The rules are in all cases explained by means of examples, in which every step of the process is clearly worked out."—*Builder.*
"One of the most serviceable books for practical mechanics. . . It is an instructive book for the student, and a text-book for him who, having once mastered the subjects it treats of, needs occasionally to refresh his memory upon them."—*Building News.*

Railway Working.

SAFE RAILWAY WORKING. A Treatise on Railway Accidents: Their Cause and Prevention; with a Description of Modern Appliances and Systems. By CLEMENT E. STRETTON, C.E., Vice-President and Consulting Engineer, Amalgamated Society of Railway Servants. With Illustrations and Coloured Plates. Third Edition, Enlarged. Crown 8vo, 3s. 6d. cloth.

"A book for the engineer, the directors, the managers; and, in short, all who wish for information on railway matters will find a perfect encyclopædia in 'Safe Railway Working.'"—*Railway Review.*
"We commend the remarks on railway signalling to all railway managers, especially where a uniform code and practice is advocated."—*Herepath's Railway Journal.*
"The author may be congratulated on having collected, in a very convenient form, much valuable information on the principal questions affecting the safe working of railways."—*Railway Engineer.*

Heat, Expansion by.

EXPANSION OF STRUCTURES BY HEAT. By JOHN KEILY, C.E., late of the Indian Public Works and Victorian Railway Departments. Crown 8vo, 3s. 6d. cloth.

SUMMARY OF CONTENTS.

Section I. FORMULAS AND DATA.
Section II. METAL BARS.
Section III. SIMPLE FRAMES.
Section IV. COMPLEX FRAMES AND PLATES.
Section V. THERMAL CONDUCTIVITY.
Section VI. MECHANICAL FORCE OF HEAT.
Section VII. WORK OF EXPANSION AND CONTRACTION.
Section VIII. SUSPENSION BRIDGES.
Section IX. MASONRY STRUCTURES.

"The aim the author has set before him, viz., to show the effects of heat upon metallic and other structures, is a laudable one, for this is a branch of physics upon which the engineer or architect can find but little reliable and comprehensive data in books."—*Builder.*
"Whoever is concerned to know the effect of changes of temperature on such structures as suspension bridges and the like, could not do better than consult Mr. Keily's valuable and handy exposition of the geometrical principles involved in these changes."—*Scotsman.*

Field Fortification.

A TREATISE ON FIELD FORTIFICATION, THE ATTACK OF FORTRESSES, MILITARY MINING, AND RECONNOITRING. By Colonel I. S. MACAULAY, late Professor of Fortification in the R.M.A., Woolwich. Sixth Edition. Crown 8vo, with separate Atlas of 12 Plates, 12s. cloth.

MR. HUMBER'S GREAT WORK ON MODERN ENGINEERING.

Complete in Four Volumes, imperial 4to, price £12 12s., half-morocco. **Each** Volume sold separately as follows:—

A RECORD OF THE PROGRESS OF MODERN ENGINEER-ING. FIRST SERIES. Comprising Civil, Mechanical, Marine, Hydraulic, Railway, Bridge, and other Engineering Works, &c. By WILLIAM HUMBER, A-M.Inst.C.E., &c. Imp. 4to, with 36 Double Plates, drawn to a large scale, Photographic Portrait of John Hawkshaw, C.E., F.R.S., &c., and copious descriptive Letterpress, Specifications, &c., £3 3s. half-morocco.

List of the Plates and Diagrams.

Victoria Station and Roof, L. B. & S. C. R. (8 plates); Southport Pier (2 plates); Victoria Station and Roof, L. C. & D. and G. W. R. (6 plates); Roof of Cremorne Music Hall; Bridge over G. N. Railway; Roof of Station, Dutch Rhenish Rail (2 plates); Bridge over the Thames, West London Extension Railway (5 plates); Armour Plates: Suspension Bridge, Thames (4 plates); The Allen Engine; Suspension Bridge, Avon (3 plates); Underground Railway (3 plates).

"Handsomely lithographed and printed. It will find favour with many who desire to preserve in a permanent form copies of the plans and specifications prepared for the guidance of the contractors for many important engineering works."—*Engineer.*

HUMBER'S PROGRESS OF MODERN ENGINEERING. SECOND SERIES. Imp. 4to, with 36 Double Plates, Photographic Portrait of Robert **Stephenson**, C.E., M.P., F.R.S., &c., and copious descriptive Letterpress, **Specifications**, &c., £3 3s. half-morocco.

List of the Plates and Diagrams.

Birkenhead Docks, Low Water Basin (15 plates); Charing Cross Station Roof, C. C. Railway (3 plates); Digswell Viaduct, Great Northern Railway; Robbery Wood Viaduct, Great Northern Railway; Iron Permanent Way; Clydach Viaduct, Merthyr, Tredegar, and Abergavenny Railway; Ebbw Viaduct, Merthyr, Tredegar, and Abergavenny Railway; College Wood Viaduct, Cornwall Railway; Dublin Winter Palace Roof (3 plates); Bridge over the Thames, L. C. & D. Railway (6 plates); Albert Harbour, Greenock (4 plates).

"Mr. Humber has done the profession good and true service, by the fine collection of examples he has here brought before the profession and the public."—*Practical Mechanic's Journal.*

HUMBER'S PROGRESS OF MODERN ENGINEERING. THIRD SERIES. Imp. 4to, with 40 Double Plates, Photographic Portrait of J. R. M'Clean, late Pres. Inst. C.E., and copious descriptive Letterpress, Specifications, &c., £3 3s. half-morocco.

List of the Plates and Diagrams.

MAIN DRAINAGE, METROPOLIS.—*North Side.*—Map showing Interception of Sewers; Middle Level Sewer (2 plates); Outfall Sewer, Bridge over River Lea (3 plates); Outfall Sewer, Bridge over Marsh Lane, North Woolwich Railway, and Bow and Barking Railway Junction; Outfall Sewer, Bridge over Bow and Barking Railway (3 plates); Outfall Sewer, Bridge over East London Waterworks' Feeder (2 plates); Outfall Sewer, Reservoir (2 plates); Outfall Sewer, Tumbling Bay and Outlet; Outfall Sewer, Penstocks. *South Side.*—Outfall Sewer, Bermondsey Branch (2 plates); Outfall Sewer, Reservoir and Outlet (4 plates); Outfall Sewer, Filth Hoist; Sections of Sewers (North and South Sides).

THAMES EMBANKMENT.—Section of River Wall; Steamboat Pier, Westminster (2 plates); Landing Stairs between Charing Cross and Waterloo Bridges; York Gate (2 plates); Overflow and Outlet at Savoy Street Sewer (3 plates); Steamboat Pier, Waterloo Bridge (3 plates); Junction of Sewers, Plans and Sections; Gullies, Plans and Sections; Rolling Stock; Granite and Iron Forts.

"The drawings have a constantly increasing value, and whoever desires to possess clear representations of the two great works carried out by our Metropolitan Board will obtain Mr. Humber's volume."—*Engineer.*

HUMBER'S PROGRESS OF MODERN ENGINEERING. FOURTH SERIES. Imp. 4to, with 36 Double Plates, Photographic Portrait of John Fowler, late Pres. Inst. C.E., and copious descriptive Letterpress Specifications, &c., £3 3s. half-morocco.

List of the Plates and Diagrams.

Abbey Mills Pumping Station, Main Drainage, Metropolis (4 plates); Barrow Docks (5 plates); Manquis Viaduct, Santiago and Valparaiso Railway (2 plates); Adam's Locomotive, St. Helen's Canal Railway (2 plates); Cannon Street Station Roof, Charing Cross Railway (3 plates); Road Bridge over the River Moka (2 plates); Telegraphic Apparatus for Mesopotamia; Viaduct over the River Wye, Midland Railway (3 plates); St. Germans Viaduct, Cornwall Railway (2 plates); Wrought-Iron Cylinder for Diving Bell; Millwall Docks (6 plates); Milroy's Patent Excavator; Metropolitan District Railway (6 plates); Harbours, Ports, and Breakwaters (3 plates).

"We gladly welcome another year's issue of this valuable publication from the able pen of Mr. Humber. The accuracy and general excellence of this work are well known, while its usefulness in giving the measurements and details of some of the latest examples of engineering, as carried out by the most eminent men in the profession, cannot be too highly prized."—*Artisan.*

THE POPULAR WORKS OF MICHAEL REYNOLDS

("THE ENGINE DRIVER'S FRIEND").

Locomotive-Engine Driving.

LOCOMOTIVE-ENGINE DRIVING : *A Practical Manual for Engineers in charge of Locomotive Engines.* By MICHAEL REYNOLDS, Member of the Society of Engineers, formerly Locomotive Inspector L. B. and S.C.R. Ninth Edition. Including a KEY TO THE LOCOMOTIVE ENGINE. With Illustrations and Portrait of Author. Crown 8vo, 4s. 6d. cloth.

"Mr. Reynolds has supplied a want, and has supplied it well. We can confidently recommend the book, not only to the practical driver, but to everyone who takes an interest in the performance of locomotive engines."—*The Engineer.*

"Mr. Reynolds has opened a new chapter in the literature of the day. Of the practical utility of this admirable treatise, we have to speak in terms of warm commendation."—*Athenæum.*

"Evidently the work of one who knows his subject thoroughly."—*Railway Service Gazette.*

"Were the cautions and rules given in the book to become part of the every-day working of our engine-drivers, we might have fewer distressing accidents to deplore."—*Scotsman.*

Stationary Engine Driving.

STATIONARY ENGINE DRIVING : *A Practical Manual for Engineers in charge of Stationary Engines.* By MICHAEL REYNOLDS. Fifth Edition, Enlarged. With Plates and Woodcuts. Crown 8vo, 4s. 6d. cloth.

"The author is thoroughly acquainted with his subjects, and his advice on the various points treated is clear and practical. . . . He has produced a manual which is an exceedingly useful one for the class for whom it is specially intended."—*Engineering.*

"Our author leaves no stone unturned. He is determined that his readers shall not only know something about the stationary engine, but all about it."—*Engineer.*

"An engineman who has mastered the contents of Mr. Reynolds's book will require but little actual experience with boilers and engines before he can be trusted to look after them."—*English Mechanic.*

The Engineer, Fireman, and Engine-Boy.

THE MODEL LOCOMOTIVE ENGINEER, FIREMAN, and ENGINE-BOY. Comprising a Historical Notice of the Pioneer Locomotive Engines and their Inventors. By MICHAEL REYNOLDS. Second Edition, with Revised Appendix. With numerous Illustrations and Portrait of George Stephenson. Crown 8vo, 4s. 6d. cloth.   [*Just published.*

"From the technical knowledge of the author it will appeal to the railway man of to-day more forcibly than anything written by Dr. Smiles. . . . The volume contains information of a technical kind, and facts that every driver should be familiar with."—*English Mechanic.*

"We should be glad to see this book in the possession of everyone in the kingdom who has ever laid, or is to lay, hands on a locomotive engine."—*Iron.*

Continuous Railway Brakes.

CONTINUOUS RAILWAY BRAKES : *A Practical Treatise on the several Systems in Use in the United Kingdom ; their Construction and Performance.* With copious Illustrations and numerous Tables. By MICHAEL REYNOLDS. Large crown 8vo, 9s. cloth.

"A popular explanation of the different brakes. It will be of great assistance in forming public opinion, and will be studied with benefit by those who take an interest in the brake."—*English Mechanic.*

"Written with sufficient technical detail to enable the principle and relative connection of the various parts of each particular brake to be readily grasped."—*Mechanical World.*

Engine-Driving Life.

ENGINE-DRIVING LIFE : *Stirring Adventures and Incidents in the Lives of Locomotive-Engine Drivers.* By MICHAEL REYNOLDS. Third and Cheaper Edition. Crown 8vo, 1s. 6d. cloth.   [*Just published.*

"From first to last perfectly fascinating. Wilkie Collins's most thrilling conceptions are thrown into the shade by true incidents, endless in their variety, related in every page."—*North British Mail.*

"Anyone who wishes to get a real insight into railway life cannot do better than read 'Engine-Driving Life' for himself ; and if he once take it up he will find that the author's enthusiasm and real love of the engine-driving profession will carry him on till he has read every page."—*Saturday Review.*

Pocket Companion for Enginemen.

THE ENGINEMAN'S POCKET COMPANION AND PRACTICAL EDUCATOR FOR ENGINEMEN, BOILER ATTENDANTS, AND MECHANICS. By MICHAEL REYNOLDS. With Forty-five Illustrations and numerous Diagrams. Third Edition, Revised. Royal 18mo, 3s. 6d., strongly bound for pocket wear.

This admirable work is well suited to accomplish its object, being the honest workmanship of a competent engineer."—*Glasgow Herald.*

"A most meritorious work, giving in a succinct and practical form all the information an engine-minder desirous of mastering the scientific principles of his daily calling would require."—*The Miller.*

"A boon to those who are striving to become efficient mechanics."—*Daily Chronicle.*

MARINE ENGINEERING, SHIPBUILDING, NAVIGATION, etc.

Pocket-Book for Naval Architects and Shipbuilders,

THE NAVAL ARCHITECT'S AND SHIPBUILDER'S POCKET-BOOK of Formulæ, Rules, and Tables, and MARINE ENGINEER'S AND SURVEYOR'S Handy Book of Reference. By CLEMENT MACKROW, Member of the Institution of Naval Architects, Naval Draughtsman. Fifth Edition, Revised and Enlarged to 700 pages, with upwards of 300 Illustrations. Fcap., 12s. 6d. strongly bound in leather.

SUMMARY OF CONTENTS.

SIGNS AND SYMBOLS, DECIMAL FRACTIONS.—TRIGONOMETRY. — PRACTICAL GEOMETRY. — MENSURATION. — CENTRES AND MOMENTS OF FIGURES.— MOMENTS OF INERTIA AND RADII OF GYRATION. — ALGEBRAICAL EXPRESSIONS FOR SIMPSON'S RULES.—MECHANICAL PRINCIPLES. — CENTRE OF GRAVITY.—LAWS OF MOTION.—DISPLACEMENT, CENTRE OF BUOYANCY.— CENTRE OF GRAVITY OF SHIP'S HULL. —STABILITY CURVES AND METACENTRES.—SEA AND SHALLOW-WATER WAVES.—ROLLING OF SHIPS.—PROPULSION AND RESISTANCE OF VESSELS. —SPEED TRIALS.—SAILING, CENTRE OF EFFORT.—DISTANCES DOWN RIVERS, COAST LINES.—STEERING AND RUDDERS OF VESSELS.—LAUNCHING CALCULATIONS AND VELOCITIES.—WEIGHT OF MATERIAL AND GEAR.—GUN PARTICULARS AND WEIGHT.—STANDARD GAUGES.—RIVETED JOINTS AND RIVETING.—STRENGTH AND TESTS OF MATERIALS. — BINDING AND SHEARING STRESSES, ETC.—STRENGTH OF SHAFTING, PILLARS, WHEELS, ETC. — HYDRAULIC DATA, ETC.—CONIC SECTIONS, CATENARIAN CURVES.—MECHANICAL POWERS, WORK. — BOARD OF TRADE REGULATIONS FOR BOILERS AND ENGINES. — BOARD OF TRADE REGULATIONS FOR SHIPS.—LLOYD'S RULES FOR BOILERS.—LLOYD'S WEIGHT OF CHAINS.—LLOYD'S SCANTLINGS FOR SHIPS.—DATA OF ENGINES AND VESSELS. · SHIPS' FITTINGS AND TESTS.— SEASONING PRESERVING TIMBER.— MEASUREMENT OF TIMBER.—ALLOYS, PAINTS, VARNISHES. — DATA FOR STOWAGE. — ADMIRALTY TRANSPORT REGULATIONS. — RULES FOR HORSEPOWER, SCREW PROPELLERS, ETC.— PERCENTAGES FOR BUTT STRAPS, ETC. —PARTICULARS OF YACHTS.—MASTING AND RIGGING VESSELS.—DISTANCES OF FOREIGN PORTS. — TONNAGE TABLES. — VOCABULARY OF FRENCH AND ENGLISH TERMS. — ENGLISH WEIGHTS AND MEASURES.—FOREIGN WEIGHTS AND MEASURES.—DECIMAL EQUIVALENTS. — FOREIGN MONEY.— DISCOUNT AND WAGE TABLES.—USEFUL NUMBERS AND READY RECKONERS —TABLES OF CIRCULAR MEASURES,— TABLES OF AREAS OF AND CIRCUMFERENCES OF CIRCLES.—TABLES OF AREAS OF SEGMENTS OF CIRCLES.— TABLES OF SQUARES AND CUBES AND ROOTS OF NUMBERS. — TABLES OF LOGARITHMS OF NUMBERS.—TABLES OF HYPERBOLIC LOGARITHMS.—TABLES OF NATURAL SINES, TANGENTS, ETC.— TABLES OF LOGARITHMIC SINES, TANGENTS, ETC.

" In these days of advanced knowledge a work like this is of the greatest value. It contains a vast amount of information. We unhesitatingly say that it is the most valuable compilation for its specific purpose that has ever been printed. No naval architect, engineer, surveyor, or seaman, wood or iron shipbuilder, can afford to be without this work."—*Nautical Magazine.*

"Should be used by all who are engaged in the construction or designs of vessels. . . . Will be found to contain the most useful tables and formulæ required by shipbuilders, carefully collected from the best authorities, and put together in a popular and simple form."—*Engineer.*

" The professional shipbuilder has now, in a convenient and accessible form, reliable data for solving many of the numerous problems that present themselves in the course of his work."—*Iron.*

" There is no doubt that a pocket-book of this description must be a necessity in the shipbuilding trade. . . . The volume contains a mass of useful information clearly expressed and presented in a handy form."—*Marine Engineer.*

Marine Engineering.

MARINE ENGINES AND STEAM VESSELS (A Treatise on). By ROBERT MURRAY, C.E. Eighth Edition, thoroughly Revised, with considerable Additions by the Author and by GEORGE CARLISLE, C.E., Senior Surveyor to the Board of Trade at Liverpool. 12mo, 5s. cloth boards.

" Well adapted to give the young steamship engineer or marine engine and boiler maker a general introduction into his practical work."—*Mechanical World.*

" We feel sure that this thoroughly revised edition will continue to be as popular in the future as it has been in the past, as, for its size, it contains more useful information than any similar treatise."—*Industries.*

" As a compendious and useful guide to engineers of our mercantile and royal naval services, we should say it cannot be surpassed."—*Building News.*

" The information given is both sound and sensible, and well qualified to direct young seagoing hands on the straight road to the extra chief's certificate. . . . Most useful to surveyors, inspectors, draughtsmen, and young engineers."—*Glasgow Herald.*

C

English-French Dictionary of Sea Terms.

TECHNICAL DICTIONARY OF SEA TERMS, PHRASES AND WORDS USED IN THE ENGLISH & FRENCH LANGUAGES. (English-French, French-English). For the Use of Seamen, Engineers, Pilots, Ship-builders, Ship-owners and Ship-brokers. Compiled by W. PIRRIE, late of the African Steamship Company. Fcap. 8vo, 5s. cloth limp.

[Just published

Pocket-Book for Marine Engineers.

A POCKET-BOOK OF USEFUL TABLES AND FORMULÆ FOR MARINE ENGINEERS. By FRANK PROCTOR, A.I.N.A. Third Edition. Royal 32mo, leather, gilt edges, with strap, 4s.

"We recommend it to our readers as going far to supply a long-felt want."—*Naval Science.*
"A most useful companion to all marine engineers."—*United Service Gazette.*

Introduction to Marine Engineering.

ELEMENTARY ENGINEERING: A Manual for Young Marine Engineers and Apprentices. In the Form of Questions and Answers on Metals, Alloys, Strength of Materials, Construction and Management of Marine Engines and Boilers, Geometry, &c. &c. With an Appendix of Useful Tables. By JOHN SHERREN BREWER, Government Marine Surveyor, Hong-kong. Second Edition, Revised. Small crown 8vo, 2s. cloth.

"Contains much valuable information for the class for whom it is intended, especially in the chapters on the management of boilers and engines."—*Nautical Magazine.*
"A useful introduction to the more elaborate text-books."—*Scotsman.*
"To a student who has the requisite desire and resolve to attain a thorough knowledge, Mr. Brewer offers decidedly useful help."—*Athenæum.*

Navigation.

PRACTICAL NAVIGATION. Consisting of THE SAILOR'S SEA-BOOK, by JAMES GREENWOOD and W. H. ROSSER; together with the requisite Mathematical and Nautical Tables for the Working of the Problems, by HENRY LAW, C.E., and Professor J. R. YOUNG. Illustrated. 12mo, 7s. strongly half-bound.

Sailmaking.

THE ART AND SCIENCE OF SAILMAKING. By SAMUEL B. SADLER, Practical Sailmaker, late in the employment of Messrs. Ratsey and Lapthorne, of Cowes and Gosport. With Plates and other Illustrations. Small 4to, 12s. 6d. cloth.

SUMMARY OF CONTENTS.

CHAP. I. THE MATERIALS USED AND THEIR RELATION TO SAILS.—II. ON THE CENTRE OF EFFORT.—III. ON MEASURING.—IV. ON DRAWING.—V. ON THE NUMBER OF CLOTHS REQUIRED.—VI. ON ALLOWANCES.—VII. CALCULATION OF GORES.—VIII. ON CUTTING OUT.—IX. ON ROPING.—X. ON DIAGONAL-CUT SAILS.—XI. CONCLUDING REMARKS.

"This work is very ably written, and is illustrated by diagrams and carefully-worked calculations. The work should be in the hands of every sailmaker, whether employer or employed, as it cannot fail to assist them in the pursuit of their important avocations."—*Isle of Wight Herald.*
"This extremely practical work gives a complete education in all the branches of the manufacture cutting out, roping, seaming, and goring. It is copiously illustrated, and will form a first-rate text-book and guide."—*Portsmouth Times.*
"The author of this work has rendered a distinct service to all interested in the art of sailmaking. The subject of which he treats is a congenial one. Mr. Sadler is a practical sailmaker, and has devoted years of careful observation and study to the subject; and the results of the experience thus gained he has set forth in the volume before us."—*Steamship.*

Chain Cables.

CHAIN CABLES AND CHAINS. Comprising Sizes and Curves of Links, Studs, &c., Iron for Cables and Chains, Chain Cable and Chain Making, Forming and Welding Links, Strength of Cables and Chains, Certificates for Cables, Marking Cables, Prices of Chain Cables and Chains, Historical Notes, Acts of Parliament, Statutory Tests, Charges for Testing, List of Manufacturers of Cables, &c. &c. By THOMAS W. TRAILL, F.E.R.N., M. Inst. C.E., Engineer Surveyor in Chief, Board of Trade, Inspector of Chain Cable and Anchor Proving Establishments, and General Superintendent, Lloyd's Committee on Proving Establishments. With numerous Tables, Illustrations and Lithographic Drawings. Folio, £2 2s. cloth.

"It contains a vast amount of valuable information. Nothing seems to be wanting to make it a complete and standard work of reference on the subject."—*Nautical Magazine.*

MINING AND METALLURGY.

Mining Machinery.

MACHINERY FOR METALLIFEROUS MINES: A Practical Treatise for Mining Engineers, Metallurgists, and Managers of Mines. By E. HENRY DAVIES, M.E., F.G.S. Crown 8vo, 580 pp., with upwards of 300 Illustrations, 12s. 6d. cloth. [*Just published.*

" Mr. Davies, in this handsome volume, has done the advanced student and the manager of mines good service. Almost every kind of machinery in actual use is carefully described, and the woodcuts and plates are good."—*Athenæum.*

" From cover to cover the work exhibits all the same characteristics which excite the confidence and attract the attention of the student as he peruses the first page. The work may safely be recommended. By its publication the literature connected with the industry will be enriched, and the reputation of its author enhanced."—*Mining Journal.*

" Mr. Davies has endeavoured to bring before his readers the best of everything in modern mining appliances. His work carries internal evidence of the author's impartiality, and this constitutes one of the great merits of the book. Throughout his work the criticisms are based on his own or other reliable experience.' —*Iron and Steel Trades' Journal.*

" The work deals with nearly every class of machinery or apparatus likely to be met with or required in connection with metalliferous mining, and is one which we have every confidence in recommending."—*Practical Engineer.*

Metalliferous Minerals and Mining.

A TREATISE ON METALLIFEROUS MINERALS AND MINING. By D. C. DAVIES, F.G.S., Mining Engineer, &c., Author of "A Treatise on Slate and Slate Quarrying." Fifth Edition, thoroughly Revised and much Enlarged, by his Son, E. HENRY DAVIES, M.E., F.G.S. With about 150 Illustrations. Crown 8vo, 12s. 6d. cloth.

"Neither the practical miner nor the general reader interested in mines can have a better book for his companion and his guide."—*Mining Journal.* [*Mining World.*

" We are doing our readers a service in calling their attention to this valuable work."

" A book that will not only be useful to the geologist, the practical miner, and the metallurgist but also very interesting to the general public."—*Iron.*

" As a history of the present state of mining throughout the world this book has a real value and it supplies an actual want."—*Athenæum.*

Earthy Minerals and Mining.

A TREATISE ON EARTHY & OTHER MINERALS AND MINING. By D. C. DAVIES, F.G.S., Author of "Metalliferous Minerals,' &c. Third Edition, revised and Enlarged, by his Son, E. HENRY DAVIES, M.E., F.G.S. With about 100 Illustrations. Crown 8vo, 12s. 6d. cloth.

" We do not remember to have met with any English work on mining matters that contains the same amount of information packed in equally convenient form."—*Academy.*

" We should be inclined to rank it as among the very best of the handy technical and trades manuals which have recently appeared."—*British Quarterly Review.*

Metalliferous Mining in the United Kingdom.

BRITISH MINING: A Treatise on the History, Discovery, Practical Development, and Future Prospects of Metalliferous Mines in the United Kingdom. By ROBERT HUNT, F.R.S., Editor of "Ure's Dictionary of Arts, Manufactures, and Mines," &c. Upwards of 950 pp., with 230 Illustrations. Second Edition, Revised. Super-royal 8vo, £2 2s. cloth.

"One of the most valuable works of reference of modern times. Mr. Hunt, as Keeper of Mining Records of the United Kingdom, has had opportunities for such a task not enjoyed by anyone else and has evidently made the most of them. . . . The language and style adopted are good, and the treatment of the various subjects laborious, conscientious, and scientific."—*Engineering.*

" The book is, in fact, a treasure-house of statistical information on mining subjects, and we know of no other work embodying so great a mass of matter of this kind. Were this the only merit of Mr. Hunt's volume, it would be sufficient to render it indispensable in the library of everyone interested in the development of the mining and metallurgical industries of this country. —*Athenæum.*

" A mass of information not elsewhere available, and of the greatest value to those who may be interested in our great mineral industries."—*Engineer.*

Underground Pumping Machinery.

MINE DRAINAGE. Being a Complete and Practical Treatise on Direct-Acting Underground Steam Pumping Machinery, with a Description of a large number of the best known Engines, their General Utility and the Special Sphere of their Action, the Mode of their Application, and their merits compared with other forms of Pumping Machinery. By STEPHEN MICHELL. 8vo, 15s. cloth.

" Will be highly esteemed by colliery owners and lessees, mining engineers, and students generally who require to be acquainted with the best means of securing the drainage of mines. It is a most valuable work, and stands almost alone in the literature of steam pumping machinery.— *Colliery Guardian.*

"Much valuable information is given, so that the book is thoroughly worthy of an extensive circulation amongst practical men and purchasers of machinery."—*Mining Journal.*

Prospecting for Gold and other Metals.

THE PROSPECTOR'S HANDBOOK: A Guide for the Prospector and Traveller in Search of Metal-Bearing or other Valuable Minerals. By J. W. ANDERSON, M.A. (Camb.), F.R.G.S., Author of "Fiji and New Caledonia," Sixth Edition, thoroughly Revised and much Enlarged. Small crown 8vo, 3s. 6d. cloth ; or, 4s. 6d. leather, pocket-book form, with tuck.

[*Just published.*

"Will supply a much felt want, especially among Colonists, in whose way are so often thrown many mineralogical specimens the value of which it is difficult to determine."—*Engineer.*

"How to find commercial minerals, and how to identify them when they are found, are the leading points to which attention is directed. The author has managed to pack as much practical detail into his pages as would supply material for a book three times its size."—*Mining Journal.*

Mining Notes and Formulæ.

NOTES AND FORMULÆ FOR MINING STUDENTS. By JOHN HERMAN MERIVALE, M.A., Certificated Colliery Manager, Professor of Mining in the Durham College of Science, Newcastle-upon-Tyne. Third Edition, Revised and Enlarged. Small crown 8vo, 2s. 6d. cloth.

"Invaluable to anyone who is working up for an examination on mining subjects."—*Iron and Coal Trades Review.*

"The author has done his work in an exceedingly creditable manner, and has produced a book that will be of service to students, and those who are practically engaged in mining operations."—*Engineer.*

Handybook for Miners.

THE MINER'S HANDBOOK : A Handy Book of Reference on the Subjects of Mineral Deposits, Mining Operations, Ore Dressing, &c. For the Use of Students and others interested in Mining matters. Compiled by JOHN MILNE, F.R.S., Professor of Mining in the Imperial University of Japan. Revised Edition. Fcap. 8vo, 7s. 6d. leather. [*Just published.*

"Professor Milne's handbook is sure to be received with favour by all connected with mining, and will be extremely popular among students."—*Athenæum.*

Miners' and Metallurgists' Pocket-Book.

A POCKET-BOOK FOR MINERS AND METALLURGISTS. Comprising Rules, Formulæ, Tables, and Notes, for Use in Field and Office Work. By F. DANVERS POWER, F.G.S., M.E. Fcap. 8vo, 9s. leather.

"This excellent book is an admirable example of its kind, and ought to find a large sale amongst English-speaking prospectors and mining engineers."—*Engineering.*

"A useful *vade-mecum* containing a mass of rules, formulæ, tables, and various other information, necessary for daily reference."—*Iron.*

Mineral Surveying and Valuing.

THE MINERAL SURVEYOR AND VALUER'S COMPLETE GUIDE, comprising a Treatise on Improved Mining Surveying and the Valuation of Mining Properties, with New Traverse Tables. By WM. LINTERN. Third Edition, Enlarged. 12mo, 4s. cloth.

"A valuable and thoroughly trustworthy guide."—*Iron and Coal Trades Review.*

Asbestos and its Uses.

ASBESTOS : Its Properties, Occurrence, and Uses. With some Account of the Mines of Italy and Canada. By ROBERT H. JONES. With Eight Collotype Plates and other Illustrations. Crown 8vo, 12s. 6d. cloth.

"An interesting and invaluable work."—*Colliery Guardian.*

Explosives.

A HANDBOOK ON MODERN EXPLOSIVES. Being a Practical Treatise on the Manufacture and Application of Dynamite, Gun-Cotton, Nitro-Glycerine, and other Explosive Compounds. Including the Manufacture of Collodion-Cotton. By M. EISSLER, Author of "The Metallurgy of Gold," &c. Crown 8vo, 10s. 6d. cloth.

"Useful not only to the miner, but also to officers of both services to whom blasting and the use of explosives generally may at any time become a necessary auxiliary."—*Nature.*

"A veritable mine of information on the subject of explosives employed for military, mining and blasting purposes."—*Army and Navy Gazette.*

Iron, Metallurgy of.

METALLURGY OF IRON. Containing History of Iron Manufacture, Methods of Assay, and Analyses of Iron Ores, Processes of Manufacture of Iron and Steel, &c. By H. BAUERMAN, F.G.S., A.R.S.M. With numerous Illustrations. Sixth Edition, Enlarged. 12mo, 5s. 6d. cloth.

Colliery Management.

THE COLLIERY MANAGER'S HANDBOOK: A Comprehensive Treatise on the Laying-out and Working of Collieries, Designed as a Book of Reference for Colliery Managers, and for the Use of Coal-Mining Students preparing for First-class Certificates. By CALEB PAMELY, Mining Engineer and Surveyor; Member of the North of England Institute of Mining and Mechanical Engineers; and Member of the South Wales Institute of Mining Engineers. With nearly 500 Plans, Diagrams, and other Illustrations. Second Edition, Revised, with Additions. Medium 8vo, about 700 pages. Price £1 5s. strongly bound.

SUMMARY OF CONTENTS.

GEOLOGY. — SEARCH FOR COAL.—MINERAL LEASES AND OTHER HOLDINGS.—SHAFT SINKING.—FITTING UP THE SHAFT AND SURFACE ARRANGEMENTS.—STEAM BOILERS AND THEIR FITTINGS.—TIMBERING AND WALLING.—NARROW WORK AND METHODS OF WORKING.—UNDERGROUND CONVEYANCE.—DRAINAGE.—THE GASES MET WITH IN MINES; VENTILATION.—ON THE FRICTION OF AIR IN MINES.— THE PRIESTMAN OIL ENGINE; PETROLEUM AND NATURAL GAS—SURVEYING AND PLANNING.—SAFETY LAMPS AND FIRE DAMP DETECTORS —SUNDRY AND INCIDENTAL OPERATIONS AND APPLIANCES.—COLLIERY EXPLOSIONS.—MISCELLANEOUS QUESTIONS & ANSWERS.

Appendix: SUMMARY OF REPORT OF H.M. COMMISSIONERS ON ACCIDENTS IN MINES.

** OPINIONS OF THE PRESS.

" Mr. Pamely has not only given us a comprehensive reference book of a very high order, suitable to the requirements of mining engineers and colliery managers, but at the same time has provided mining students with a class-book that is as interesting as it is instructive."—*Colliery Manager.*

" Mr. Pamely's work is eminently suited to the purpose for which it is intended—being clear, interesting, exhaustive, rich in detail, and up to date, giving descriptions of the very latest machines in every department. . . . A mining engineer could scarcely go wrong who followed this work."—*Colliery Guardian.*

" This is the most complete 'all round' work on coal-mining published in the English language. . . . No library of coal-mining books is complete without it."—*Colliery Engineer* (Scranton, Pa., U.S.A.).

" Mr. Pamely's work is in all respects worthy of our admiration. No person in any responsible position connected with mines should be without a copy."—*Westminster Review.*

Coal and Iron.

THE COAL AND IRON INDUSTRIES OF THE UNITED KINGDOM. Comprising a Description of the Coal Fields, and of the Principal Seams of Coal, with Returns of their Produce and its Distribution, and Analyses of Special Varieties. Also an Account of the occurrence of Iron Ores in Veins or Seams; Analyses of each Variety; and a History of the Rise and Progress of Pig Iron Manufacture. By RICHARD MEADE, Assistant Keeper of Mining Records. With Maps. 8vo, £1 8s. cloth.

"The book is one which must find a place on the shelves of all interested in coal and iron production, and in the iron, steel, and other metallurgical industries."—*Engineer.*

"Of this book we may unreservedly say that it is the best of its class which we have ever met. . . . A book of reference which no one engaged in the iron or coal trades should omit from his library."—*Iron and Coal Trades Review.*

Coal Mining.

COAL AND COAL MINING: A Rudimentary Treatise on. By the late Sir WARINGTON W. SMYTH, M.A., F.R.S., &c., Chief Inspector of the Mines of the Crown. Seventh Edition, Revised and Enlarged. With numerous Illustrations. 12mo, 4s. cloth boards.

" As an outline is given of every known coal-field in this and other countries, as well as of the principal methods of working, the book will doubtless interest a very large number of readers."— *Mining Journal.*

Subterraneous Surveying.

SUBTERRANEOUS SURVEYING, Elementary and Practical Treatise on, with and without the Magnetic Needle. By THOMAS FENWICK, Surveyor of Mines, and THOMAS BAKER, C.E. Illust. 12mo, 3s. cloth boards.

Granite Quarrying.

GRANITES AND OUR GRANITE INDUSTRIES. By GEORGE F. HARRIS. F.G.S., Membre de la Société Belge de Géologie, Lecturer on Economic Geology at the Birkbeck Institution, &c. With Illustrations. Crown 8vo, 2s. 6d., cloth.

" A clearly and well-written manual on the granite industry." —*Scotsman.*

" An interesting work, which will be deservedly esteemed."—*Colliery Guardian.*

" An exceedingly interesting and valuable monograph on a subject which has hitherto received unaccountably little attention in the shape of systematic literary treatment."—*Scottish Leader.*

Gold, Metallurgy of.

THE METALLURGY OF GOLD: A Practical Treatise on the Metallurgical Treatment of Gold-bearing Ores. Including the Processes of Concentration and Chlorination, and the Assaying, Melting, and Refining of Gold. By M. EISSLER, Mining Engineer and Metallurgical Chemist, formerly Assistant Assayer of the U. S. Mint, San Francisco. Third Edition, Revised and greatly Enlarged. With 187 Illustrations. Crown 8vo, 12s. 6d. cloth.

" This book thoroughly deserves its title of a 'Practical Treatise.' The whole process of gold milling, from the breaking of the quartz to the assay of the bullion, is described in clear and orderly narrative and with much, but not too much, fulness of detail."—*Saturday Review.*

" The work is a storehouse of information and valuable data, and we strongly recommend it to all professional men engaged in the gold-mining industry."—*Mining Journal.*

Gold Extraction.

THE CYANIDE PROCESS OF GOLD EXTRACTION: and its Practical Application on the Witwatersrand Gold Fields in South Africa. By M. EISSLER, M.E., Mem. Inst. Mining and Metallurgy, Author of "The Metallurgy of Gold," &c. With Diagrams and Working Drawings. Large crown 8vo, 7s. 6d. cloth.   [*Just published.*

" This book is just what was needed to acquaint mining men with the actual working of a process which is not only the most popular, but is, as a general rule, the most successful for the extraction of gold from tailings."—*Mining Journal.*

" The work will prove invaluable to all interested in gold mining, whether metallurgists or as investors."—*Chemical News.*

Silver, Metallurgy of.

THE METALLURGY OF SILVER: A Practical Treatise on the Amalgamation, Roasting, and Lixiviation of Silver Ores. Including the Assaying, Melting and Refining, of Silver Bullion. By M. EISSLER, Author of "The Metallurgy of Gold," &c. Second Edition, Enlarged. With 150 Illustrations. Crown 8vo, 10s. 6d. cloth.

" A practical treatise, and a technical work which we are convinced will supply a long-felt want amongst practical men, and at the same time be of value to students and others indirectly connected with the industries."—*Mining Journal.*

" From first to last the book is thoroughly sound and reliable."—*Colliery Guardian.*

" For chemists, practical miners, assayers, and investors alike, we do not know of any work on the subject so handy and yet so comprehensive."—*Glasgow Herald.*

Lead, Metallurgy of.

THE METALLURGY OF ARGENTIFEROUS LEAD: A Practical Treatise on the Smelting of Silver-Lead Ores and the Refining of Lead Bullion. Including Reports on various Smelting Establishments and Descriptions of Modern Smelting Furnaces and Plants in Europe and America. By M. EISSLER, M.E., Author of "The Metallurgy of Gold," &c. Crown 8vo, 400 pp., with 183 Illustrations, 12s. 6d. cloth.

" **The** numerous metallurgical processes, which are fully and extensively treated of, embrace all the stages experienced in the passage of the lead from the various natural states to its issue from the refinery **as an** article of commerce."—*Practical Engineer.*

" The present volume fully maintains the reputation of the author. Those who wish to obtain a thorough insight into the present state of this industry cannot do better than read this volume, and all mining engineers cannot fail to find many useful hints and suggestions in it."—*Industries.*

" It is most carefully written and illustrated **with** capital drawings and diagrams. In fact, it **is** the work of an expert for experts, by whom **it will be** prized as an indispensable text-book."—*Bristol Mercury.*

Iron Mining.

THE IRON ORES OF GREAT BRITAIN AND IRELAND: Their Mode of Occurrence, Age, and Origin, and the Methods of Searching for and Working them, with a Notice of some of the Iron Ores of Spain. By J. D. KENDALL, F.G.S., Mining Engineer. Crown 8vo, 16s. cloth.

" The author has a thorough practical knowledge of his subject, and has supplemented a careful study of the available literature by unpublished information derived from his own observations. The result is a very useful volume which cannot fail to be of value to all interested in the iron industry of the country."—*Industries.*

" Mr. Kendall is a great authority on this subject and writes rom personal observation."—*Colliery Guardian.*

" Mr. Kendall's book is thoroughly well done. In it there are the outlines of the history of ore mining in every centre and there is everything that we want to know as to the character of the ores of each district, their commercial value and the cost of working them "—*Iron and Steel Trades Journal.*

ELECTRICITY, ELECTRICAL ENGINEERING, etc.

Dynamo Management.

THE MANAGEMENT OF DYNAMOS: A Handybook of Theory and Practice for the Use of Mechanics, Engineers, Students and others in Charge of Dynamos. By G. W. LUMMIS PATERSON. With numerous Illustrations. Crown 8vo, 3s. 6d. cloth. *[Just published.*

Electrical Engineering.

THE ELECTRICAL ENGINEER'S POCKET-BOOK OF MODERN RULES, FORMULÆ, TABLES, AND DATA. By H. R. KEMPE, M.Inst.E.E., A.M.Inst.C.E., Technical Officer, Postal Telegraphs, Author of "A Handbook of Electrical Testing," &c. Second Edition, thoroughly Revised, with Additions. Royal 32mo, oblong, 5s. leather.

"There is very little in the shape of formulæ or data which the electrician is likely to want in a hurry which cannot be found in its pages."—*Practical Engineer.*

"A very useful book of reference for daily use in practical electrical engineering and its various applications to the industries of the present day."—*Iron.*

"It is the best book of its kind."—*Electrical Engineer.*

"Well arranged and compact. The 'Electrical Engineer's Pocket-Book' is a good one."—*Electrician.* [*Review.*

"Strongly recommended to those engaged in the various electrical industries."—*Electrical*

Electric Lighting.

ELECTRIC LIGHT FITTING: A Handbook for Working Electrical Engineers, embodying Practical Notes on Installation Management. By JOHN W. URQUHART, Electrician, Author of "Electric Light." &c. With numerous Illustrations. Second Edition, Revised, with Additional Chapters. Crown 8vo, 5s. cloth.

"This volume deals with what may be termed the mechanics of electric lighting, and is addressed to men who are already engaged in the work or are training for it. The work traverses a great deal of ground, and may be read as a sequel to the same author's useful work on 'Electric Light.'"—*Electrician.*

"This is an attempt to state in the simplest language the precautions which should be adopted in installing the electric light, and to give information, for the guidance of those who have to run the plant when installed. The book is well worth the perusal of the workmen for whom it is written."—*Electrical Review.*

"We have read this book with a good deal of pleasure. We believe that the book will be of use to practical workmen, who will not be alarmed by finding mathematical formulæ which they are unable to understand."—*Electrical Plant.*

Electric Light.

ELECTRIC LIGHT: Its Production and Use. Embodying Plain Directions for the Treatment of Dynamo-Electric Machines, Batteries Accumulators, and Electric Lamps. By J. W. URQUHART, C.E., Author of "Electric Light Fitting," "Electroplating," &c. Fifth Edition, carefully Revised, with Large Additions and 145 Illustrations. Crown 8vo, 7s. 6d. cloth.

"The whole ground of electric lighting is more or less covered and explained in a very clear and concise manner."—*Electrical Review.*

"Contains a good deal of very interesting information, especially in the parts where the author gives dimensions and working costs."—*Electrical Engineer.*

"A miniature *vade-mecum* of the salient facts connected with the science of electric lighting."—*Electrician.*

"You cannot for your purpose have a better book than 'Electric Light,' by Urquhart."—*Engineer.*

"The book is by far the best that we have yet met with on the subject."—*Athenæum.*

Construction of Dynamos.

DYNAMO CONSTRUCTION: A Practical Handbook for the Use of Engineer Constructors and Electricians-in-Charge. Embracing Framework Building, Field Magnet and Armature Winding and Grouping, Compounding, &c. With Examples of leading English, American, and Continental Dynamos and Motors. By J. W. URQUHART, Author of "Electric Light," "Electric Light Fitting," &c. Second Edition, Revised and Enlarged. With 114 Illustrations. Crown 8vo, 7s. 6d. cloth. [*Just published.*

"Mr. Urquhart's book is the first one which deals with these matters in such a way that the engineering student can understand them. The book is very readable, and the author leads his readers up to difficult subjects by reasonably simple tests."—*Engineering Review.*

"The author deals with his subject in a style so popular as to make his volume a handbook of great practical value to engineer constructors and electricians in charge."—*Scotsman.*

"'Dynamo Construction' more than sustains the high character of the author's previous publications. It is sure to be widely read by the large and rapidly increasing number of practical electricians."—*Glasgow Herald.*

"A book for which a demand has long existed."—*Mechanical World.*

A New Dictionary of Electricity.

THE STANDARD ELECTRICAL DICTIONARY. A Popular Dictionary of Words and Terms Used in the Practice of Electrical Engineering. Containing upwards of 3,000 Definitions. By T. O'CONNOR SLOANE, A.M., Ph.D., Author of "The Arithmetic of Electricity," &c. Crown 8vo, 6so pp., 350 Illustrations, 7s. 6d. cloth.   [*Just published.*

"The work has many attractive features in it, and is beyond doubt, a well put together and useful publication. The amount of ground covered may be gathered from the fact that in the index about 5,000 references will be found. The inclusion of such comparatively modern words as 'impedence,' 'reluctance,' &c., shows that the author has desired to be up to date, and indeed there are other indications of carefulness of compilation. The work is one which does the author great credit and it should prove of great value, especially to students."—*Electrical Review.*

"Very complete and contains a large amount of useful information."—*Industries.*

"An encyclopædia of electrical science in the compass of a dictionary. The information given is sound and clear. The book is well printed, well illustrated, and well up to date, and may be confidently recommended."—*Builder.*

"The volume is excellently printed and illustrated, and should form part of the library of every one who is connected with electrical matters."—*Hardware Trade Journal.*

Electric Lighting of Ships.

ELECTRIC SHIP-LIGHTING : A Handbook on the Practical Fitting and Running of Ship's Electrical Plant. For the Use of Shipowners and Builders, Marine Electricians, and Sea-going Engineers-in-Charge. By J. W. URQUHART, C.E. With 88 Illustrations. Crown 8vo, 7s 6d. cloth.

"The subject of ship electric lighting is one of vast importance in these days, and Mr. Urquhart is to be highly complimented for placing such a valuable work at the service of the practical marine electrician."—*The Steamship.*

"Distinctly a book which of its kind stands almost alone, and for which there should be a demand."—*Electrical Review.*

Country House Electric Lighting.

ELECTRIC LIGHT FOR COUNTRY HOUSES : A Practical Handbook on the Erection and Running of Small Installations, with particulars of the Cost of Plant and Working. By J. H. KNIGHT. Crown 8vo, 1s. wrapper.   [*Just published.*

Electric Lighting.

THE ELEMENTARY PRINCIPLES OF ELECTRIC LIGHTING. By ALAN A. CAMPBELL SWINTON, Associate I.E.E. Third Edition, Enlarged and Revised. With 16 Illustrations. Crown 8vo, 1s. 6d. cloth.

"Anyone who desires a short and thoroughly clear exposition of the elementary principles of electric-lighting cannot do better than read this little work."—*Bradford Observer.*

Dynamic Electricity.

THE ELEMENTS OF DYNAMIC ELECTRICITY AND MAGNETISM. By PHILIP ATKINSON, A.M., Ph.D., Author of "The Elements of Electric Lighting," &c. Cr. 8vo, with 120 Illustrations, 10s. 6d. cl.

Electric Motors, &c.

THE ELECTRIC TRANSFORMATION OF POWER and its Application by the Electric Motor, including Electric Railway Construction. By P. ATKINSON, A.M., Ph.D, Author of "The Elements of Electric Lighting," &c. With 96 Illustrations. Crown 8vo, 7s. 6d. cloth.

Dynamo Construction.

HOW TO MAKE A DYNAMO: A Practical Treatise for Amateurs. Containing numerous Illustrations and Detailed Instructions for Constructing a Small Dynamo, to Produce the Electric Light. By ALFRED CROFTS. Fourth Edition, Revised and Enlarged. Crown 8vo, 2s. cloth,

"The instructions given in this unpretentious little book are sufficiently clear and explicit to enable any amateur mechanic possessed of average skill and the usual tools to be found in an amateur's workshop, to build a practical dynamo machine."—*Electrician.*

Text Book of Electricity.

THE STUDENT'S TEXT-BOOK OF ELECTRICITY. By HENRY M. NOAD, F.R.S. 630 pages, with 470 Illustrations. Cheaper Edition. Crown 8vo, 9s. cloth.   [*Just published.*

Electricity.

A MANUAL OF ELECTRICITY : Including Galvanism, Magnetism, Dia-Magnetism, Electro-Dynamics. By HENRY M. NOAD, Ph D., F.R.S. Fourth Edition (1859). 8vo, £1 4s. cloth.

ARCHITECTURE, BUILDING, etc.

Building Construction.

PRACTICAL BUILDING CONSTRUCTION: A Handbook for Students Preparing for Examinations, and a Book of Reference for Persons Engaged in Building. By JOHN P. ALLEN, Surveyor, Lecturer on Building Construction at the Durham College of Science, Newcastle. Medium 8vo, 450 pages, with 1,000 Illustrations. 12s. 6d. cloth.   [*Just published.*

" This volume is one of the most complete expositions of building construction we have seen. It contains all that is necessary to prepare students for the various examinations in building construction."—*Building News.*

" The author depends nearly as much **on his** diagrams **as on his** type. The pages suggest the hand of a man of experience in building operations—and **the volume must be** a blessing to many teachers as well as to students."—*The Architect.*

" The work is sure to prove a formidable rival to great and small competitors alike, and bids fair to take a permanent place as a favourite students' text-book. The large number of illustrations deserve particular mention for the great merit they possess for purposes of reference, in exactly corresponding to convenient scales."—*Jour. Inst. Brit. Archts.*

The New London Building Act, 1894.

THE LONDON BUILDING ACT, 1894; with the By-Laws and Regulations of the London County Council, and Introduction, Notes, Cases and Index. By ALEX. J. DAVID, B.A., LL.M. of the Inner Temple, Barrister-at-Law. Crown 8vo, 3s. 6d. cloth.   [*Just published.*

" To all architects and district surveyors and builders, Mr. David's manual will be welcome."—*Building News.*

" The volume will doubtless **be eagerly** consulted by **the building** fraternity."—*Illustrated Carpenter and Builder.*

Concrete.

CONCRETE : ITS NATURE AND USES. A Book for Architects, Builders, Contractors, and Clerks of Works. By GEORGE L. SUTCLIFFE, A.R.I.B.A. Crown 8vo, 7s. 6d. cloth.   [*Just published.*

" The author treats a difficult subject in a lucid manner. The manual fills a long-felt gap. It is careful and exhaustive ; equally useful as a student's guide and a architect's book of reference."—*Journal of Royal Institution of British Architects.*

" There is room for this new book, which will probably be for some time the standard work on the subject for a builder's purpose."—*Glasgow Herald.*

" A thoroughly useful and comprehensive work."—*British Architect.*

Mechanics for Architects.

THE MECHANICS OF ARCHITECTURE : A Treatise on Applied Mechanics, especially Adapted to the Use of Architects. By E. W. TARN, M.A., Author of "The Science of Building," &c. Second Edition, Enlarged. Illust. with 125 Diagrams. Cr. 8vo, 7s. 6d. cloth. [*Just published.*

" The book is a very useful and helpful manual of architectural mechanics, and really contains sufficient to enable **a** careful and painstaking student to grasp the principles bearing upon the majority of building problems. . . . Mr. Tarn has added, by this volume, to the debt of gratitude which is owing to him by architectural students for the many valuable works which he has produced for their use."—*The Builder.*

" The mechanics in the volume are really mechanics, and are harmoniously wrought in with the distinctive professional manner proper to the subject."—*The Schoolmaster.*

The New Builder's Price Book, 1896.

LOCKWOOD'S BUILDER'S PRICE BOOK FOR 1896. A Comprehensive Handbook of the Latest Prices and Data for Builders, Architects, Engineers, and Contractors. By FRANCIS T. W. MILLER. 800 closely-printed pages, crown 8vo, 4s. cloth.

" This book is a very useful one, and should find a place in every English office connected with the building and engineering professions."—*Industries.*

" An excellent book of reference."—*Architect.*

" In its new and revised form **this** Price Book is what a work of this kind should be—comprehensive, reliable, well arranged, legible, and well bound."—*British Architect.*

Designing Buildings.

THE DESIGN OF BUILDINGS: Being Elementary Notes on the Planning, Sanitation and Ornamentive Formation of Structures, based on Modern Practice. Illustrated with Nine Folding Plates. By W. WOODLEY, Assistant Master, Metropolitan Drawing Classes, &c. 8vo, 6s. cloth.

Sir Wm. Chambers's Treatise on Civil Architecture.

THE DECORATIVE PART OF CIVIL ARCHITECTURE. By Sir WILLIAM CHAMBERS, F.R.S. With Portrait, Illustrations, Notes, and an Examination of Grecian Architecture, by JOSEPH GWILT, F.S.A. Revised and Edited by W. H. LEEDS. 66 Plates, 4to, 21s. cloth.

Villa Architecture.

A HANDY BOOK OF VILLA ARCHITECTURE: Being a Series of Designs for Villa Residences in various Styles. With Outline Specifications and Estimates. By C. WICKES, Architect, Author of "The Spires and Towers of England," &c. 61 Plates, 4to, £1 11s. 6d. half-morocco.

"The whole of the designs bear evidence of their being the work of an artistic architect, and they will prove very valuable and suggestive."—*Building News.*

Text-Book for Architects.

THE ARCHITECT'S GUIDE: Being a Text-Book of Useful Information for Architects, Engineers, Surveyors, Contractors, Clerks of Works, &c. &c. By FREDERICK ROGERS, Architect. Third Edition. Crown 8vo, 3s. 6d. cloth.

"As a text-book of useful information for architects, engineers, surveyors, &c., it would be hard to find a handier or more complete little volume."—*Standard.*

Taylor and Cresy's Rome.

THE ARCHITECTURAL ANTIQUITIES OF ROME. By the late G. L. TAYLOR, Esq., F.R.I.B.A., and EDWARD CRESY, Esq. New Edition, thoroughly Revised by the Rev. ALEXANDER TAYLOR, M.A. (son of the late G. L. Taylor, Esq.), Fellow of Queen's College, Oxford, and Chaplain of Gray's Inn. Large folio, with 130 Plates, £3 3s. half-bound.

"Taylor and Cresy's work has from its first publication been ranked among those professional books which cannot be bettered."—*Architect.*

Linear Perspective.

ARCHITECTURAL PERSPECTIVE: The whole Course and Operations of the Draughtsman in Drawing a Large House in Linear Perspective. Illustrated by 43 Folding Plates. By F. O. FERGUSON. Second Edition, Enlarged. 8vo, 3s. 6d. boards.   [*Just published.*]

"It is the most intelligible of the treatises on this ill-treated subject that I have met with."—E. INGRESS BELL, Esq., in the *R.I.B.A. Journal.*

Architectural Drawing.

PRACTICAL RULES ON DRAWING, for the Operative Builder and Young Student in Architecture. By GEORGE PYNE. With 14 Plates, 4to, 7s. 6d. boards.

Vitruvius' Architecture.

THE ARCHITECTURE of MARCUS VITRUVIUS POLLIO. Translated by JOSEPH GWILT, F.S.A., F.R.A.S. New Edition, Revised by the Translator. With 23 Plates. Fcap. 8vo, 5s. cloth.

Designing, Measuring, and Valuing.

THE STUDENT'S GUIDE to the PRACTICE of MEASURING AND VALUING ARTIFICERS' WORK. Containing Directions for taking Dimensions, Abstracting the same, and bringing the Quantities into Bill, with Tables of Constants for Valuation of Labour, and for the Calculation of Areas and Solidities. Originally edited by EDWARD DOBSON, Architect. With Additions by E. WYNDHAM TARN, M.A. Sixth Edition. With 8 Plates and 63 Woodcuts. Crown 8vo, 7s. 6d. cloth.

"This edition will be found the most complete treatise on the principles of measuring and valuing artificers' work that has yet been published."—*Building News.*

Pocket Estimator and Technical Guide.

THE POCKET TECHNICAL GUIDE, MEASURER, AND ESTIMATOR FOR BUILDERS AND SURVEYORS. Containing Technical Directions for Measuring Work in all the Building Trades, Complete Specifications for Houses, Roads, and Drains, and an easy Method of Estimating the parts of a Building collectively. By A. C. BEATON. Seventh Edit. Waistcoat-pocket size, 1s. 6d. leather, gilt edges.

"No builder, architect, surveyor, or valuer should be without his 'Beaton.'"—*Building News.*

Donaldson on Specifications.

THE HANDBOOK OF SPECIFICATIONS; or, Practical Guide to the Architect, Engineer, Surveyor, and Builder, in drawing up Specifications and Contracts for Works and Constructions. Illustrated by Precedents of Buildings actually executed by eminent Architects and Engineers. By Professor T. L. DONALDSON, P.R.I.B.A., &c. New Edition. 8vo, with upwards of 1,000 pages of Text, and 33 Plates, £1 11s. 6d. cloth.

"Valuable as a record, and more valuable still as a book of precedents. . . . Suffice it to say that Donaldson's 'Handbook of Specifications' must be bought by all architects."—*Builder*

Bartholomew and Rogers' Specifications.

SPECIFICATIONS FOR PRACTICAL ARCHITECTURE. A Guide to the Architect, Engineer, Surveyor, and Builder. With an Essay on the Structure and Science of Modern Buildings. Upon the Basis of the Work by ALFRED BARTHOLOMEW, thoroughly Revised, Corrected, and greatly added to by FREDERICK ROGERS, Architect. Third Edition, Revised, with Additions. With numerous Illustrations. Medium 8vo, 15s. cloth.

"The collection of specifications prepared by Mr. Rogers on the basis of Bartholomew's work is too well known to need any recommendation from us. It is one of the books with which every young architect must be equipped."—*Architect.*

Construction.

THE SCIENCE OF BUILDING : An Elementary Treatise on the Principles of Construction. By E. WYNDHAM TARN, M.A., Architect. Third Edition, Revised and Enlarged. With 59 Engravings. Fcap. 8vo, 4s. cl.

"A very valuable book, which we strongly recommend to all students."—*Builder.*

House Building and Repairing.

THE HOUSE-OWNER'S ESTIMATOR ; or, What will it Cost to Build, Alter, or Repair? A Price Book for Unprofessional People, as well as the Architectural Surveyor and Builder. By J. D. SIMON. Edited by F. T. W. MILLER, A.R.I.B.A. Fourth Edition. Crown 8vo, 3s. 6d. cloth.

"In two years it will repay its cost a hundred times over."—*Field.*

Cottages and Villas.

COUNTRY AND SUBURBAN COTTAGES AND VILLAS : How to Plan and Build Them. Containing 33 Plates, with Introduction, General Explanations, and Description of each Plate. By JAMES W. BOGUE, Architect, Author of "Domestic Architecture," &c. 4to, 10s. 6d. cloth.

Building ; Civil and Ecclesiastical.

A BOOK ON BUILDING, Civil and Ecclesiastical, including Church Restoration ; with the Theory of Domes and the Great Pyramid, &c. By Sir EDMUND BECKETT, Bart., LL.D., F.R.A.S. Fcap. 8vo, 5s. cloth.

"A book which is always amusing and nearly always instructive."—*Times.*

Sanitary Houses, etc.

THE SANITARY ARRANGEMENT OF DWELLING-HOUSES: A Handbook for Householders and Owners of Houses. By A. J. WALLIS-TAYLER, A.M. Inst. C.E. With numerous Illustrations. Crown 8vo, 2s. 6d. cloth.   [*Just published.*

"This book will be largely read ; it will be of considerable service to the public. It is well arranged, easily read, and for the most part devoid of technical terms."—*Lancet.*

Ventilation of Buildings.

VENTILATION. A Text Book to the Practice of the Art of Ventilating Buildings. By W. P. BUCHAN, R.P. 12mo, 4s. cloth.

"Contains a great amount of useful practical information, as thoroughly interesting as it is technically reliable."—*British Architect.*

The Art of Plumbing.

PLUMBING. A Text Book to the Practice of the Art or Craft of the Plumber. By WILLIAM PATON BUCHAN, R.P. Sixth Edition. 4s. cloth.

"A text-book which may be safely put in the hands of every young plumber."—*Builder.*

Geometry for the Architect, Engineer, etc.

PRACTICAL GEOMETRY, for the Architect, Engineer, and Mechanic. Giving Rules for the Delineation and Application of various Geometrical Lines, Figures and Curves. By E. W. TARN, M.A., Architect. 8vo, 9s. cloth.

"No book with the same objects in view has ever been published in which the clearness of the rules laid down and the illustrative diagrams have been so satisfactory."—*Scotsman.*

The Science of Geometry.

THE GEOMETRY OF COMPASSES; or, Problems Resolved by the mere Description of Circles, and the use of Coloured Diagrams and Symbols. By OLIVER BYRNE. Coloured Plates. Crown 8vo, 3s. 6d. cloth.

CARPENTRY, TIMBER, etc.

Tredgold's Carpentry, Revised & Enlarged by Tarn.

THE ELEMENTARY PRINCIPLES OF CARPENTRY. A Treatise on the Pressure and Equilibrium of Timber Framing, the Resistance of Timber, and the Construction of Floors, Arches, Bridges, Roofs, Uniting Iron and Stone with Timber, &c. To which is added an Essay on the Nature and Properties of Timber, &c., with Descriptions of the kinds of Wood used in Building; also numerous Tables of the Scantlings of Timber for different purposes, the Specific Gravities of Materials, &c. By THOMAS TREDGOLD, C.E. With an Appendix of Specimens of Various Roofs of Iron and Stone, Illustrated. Seventh Edition, thoroughly revised and considerably enlarged by E. WYNDHAM TARN, M.A., Author of "The Science of Building," &c. With 61 Plates, Portrait of the Author, and several Woodcuts. In One large Vol., 4to, price £1 5s. cloth.

"Ought to be in every architect's and every builder's library."—*Builder.*
"A work whose monumental excellence must commend it wherever skilful carpentry is concerned. The author's principles are rather confirmed than impaired by time. The additional plates are of great intrinsic value."—*Building News.*

Woodworking Machinery.

WOODWORKING MACHINERY: Its Rise, Progress, and Construction. With Hints on the Management of Saw Mills and the Economical Conversion of Timber. Illustrated with Examples of Recent Designs by leading English, French, and American Engineers. By M. POWIS BALE, A.M.Inst.C.E., M.I.M.E. Second Edition, Revised, with large Additions. Large crown 8vo, 440 pp., 9s. cloth.   [*Just published.*

"Mr. Bale is evidently an expert on the subject and he has collected so much information that the book is all-sufficient for builders and others engaged in the conversion of timber."—*Architect.*
"The most comprehensive compendium of wood-working machinery we have seen. The author is a thorough master of his subject."—*Building News.*

Saw Mills.

SAW MILLS: Their Arrangement and Management, and the Economical Conversion of Timber. (A Companion Volume to "Woodworking Machinery.") By M. POWIS BALE. Crown 8vo, 10s. 6d. cloth.

"The *administration* of a large sawing establishment is discussed, and the subject examined from a financial standpoint. Hence the size, shape, order, and disposition of saw-mills and the like are gone into in detail, and the course of the timber is traced from its reception to its delivery in its converted state. We could not desire a more complete or practical treatise."—*Builder.*

Nicholson's Carpentry.

THE CARPENTER'S NEW GUIDE: or, Book of Lines for Carpenters; comprising all the Elementary Principles essential for acquiring a knowledge of Carpentry. Founded on the late PETER NICHOLSON'S Standard Work. New Edition, Revised by A. ASHPITEL, F.S.A. With Practical Rules on Drawing, by G. PYNE. With 74 Plates, 4to, £1 1s. cloth.

Handrailing and Stairbuilding.

A PRACTICAL TREATISE ON HANDRAILING: Showing New and Simple Methods for Finding the Pitch of the Plank, Drawing the Moulds, Bevelling, Jointing-up, and Squaring the Wreath. By GEORGE COLLINGS. Second Edition, Revised and Enlarged, to which is added A TREATISE ON STAIRBUILDING. 12mo, 2s. 6d. cloth limp.

"Will be found of practical utility in the execution of this difficult branch of joinery."—*Builder.*
"Almost every difficult phase of this somewhat intricate branch of joinery is elucidated by the aid of plates and explanatory letterpress."—*Furniture Gazette.*

Circular Work.

CIRCULAR WORK IN CARPENTRY AND JOINERY: A Practical Treatise on Circular Work of Single and Double Curvature. By GEORGE COLLINGS. With Diagrams. Second Edit. 12mo, 2s. 6d. cloth limp.

"An excellent example of what a book of this kind should be. Cheap in price, clear in definition and practical in the examples selected."—*Builder.*

Handrailing.

HANDRAILING COMPLETE IN EIGHT LESSONS. On the Square-Cut System. By J. S. GOLDTHORP, Teacher of Geometry and Building Construction at the Halifax Mechanic's Institute. With Eight Plates and over 150 Practical Exercises. 4to, 3s. 6d. cloth.

"Likely to be of considerable value to joiners and others who take a pride in good work. We heartily commend it to teachers and students."—*Timber Trades Journal.*

Timber Merchant's Companion.

THE TIMBER MERCHANT'S AND BUILDER'S COM-
PANION. Containing New and Copious Tables of the Reduced Weight and
Measurement of Deals and Battens, of all sizes, from One to a Thousand
Pieces, and the relative Price that each size bears per Lineal Foot to any
given Price per Petersburg Standard Hundred; the Price per Cube Foot of
Square Timber to any given Price per Load of 50 Feet; the proportionate
Value of Deals and Battens by the Standard, to Square Timber by the Load
of 50 Feet; the readiest mode of ascertaining the Price of Scantling per
Lineal Foot of any size, to any given Figure per Cube Foot, &c. &c. By
WILLIAM DOWSING. Fourth Edition, Revised and Corrected. Cr. 8vo, 3s. cl.
"Everything is as concise and clear as it can possibly be made. There can be no doubt that
every timber merchant and builder ought to possess it."—*Hull Advertiser.*
"We are glad to see a fourth edition of these admirable tables, which for correctness and
simplicity of arrangement leave nothing to be desired."—*Timber Trades Journal.*

Practical Timber Merchant.

THE PRACTICAL TIMBER MERCHANT. Being a Guide
for the use of Building Contractors, Surveyors, Builders, &c., comprising
useful Tables for all purposes connected with the Timber Trade, Marks of
Wood, Essay on the Strength of Timber, Remarks on the Growth of Timber,
&c. By W. RICHARDSON. Second Edition. Fcap. 8vo, 3s. 6d. cloth.
"This handy manual contains much valuable information for the use of timber merchants,
builders, foresters, and all others connected with the growth, sale, and manufacture of timber."—
Journal of Forestry.

Packing-Case Makers, Tables for.

PACKING-CASE TABLES; showing the number of Super-
ficial Feet in Boxes or Packing-Cases, from six inches square and upwards.
By W. RICHARDSON, Timber Broker. Third Edition. Oblong 4to, 3s. 6d. cl.
"Invaluable labour-saving tables."—*Ironmonger.*
"Will save much labour and calculation."—*Grocer.*

Superficial Measurement.

THE TRADESMAN'S GUIDE TO SUPERFICIAL MEA-
SUREMENT. Tables calculated from 1 to 200 inches in length, by 1 to 108
inches in breadth. For the use of Architects, Surveyors, Engineers, Timber
Merchants, Builders, &c. By JAMES HAWKINGS. Fourth Edition. Fcap.,
3s. 6d. cloth.
"A useful collection of tables to facilitate rapid calculation of surfaces. The exact area of any
surface of which the limits have been ascertained can be instantly determined. The book will be
found of the greatest utility to all engaged in building operations."—*Scotsman.*
"These tables will be found of great assistance to all who require to make calculations in superficial measurement."—*English Mechanic.*

Forestry.

THE ELEMENTS OF FORESTRY. Designed to afford In-
formation concerning the Planting and Care of Forest Trees for Ornament or
Profit, with Suggestions upon the Creation and Care of Woodlands. By F. B.
HOUGH. Large crown 8vo, 10s. cloth.

Timber Importer's Guide.

THE TIMBER IMPORTER'S, TIMBER MERCHANT'S, AND
BUILDER'S STANDARD GUIDE. By RICHARD E. GRANDY. Compris-
ing an Analysis of Deal Standards, Home and Foreign, with Comparative
Values and Tabular Arrangements for fixing Net Landed Cost on Baltic
and North American Deals, including all intermediate Expenses, Freight,
Insurance, &c. &c. Together with copious Information for the Retailer and
Builder. Third Edition, Revised. 12mo, 2s. cloth limp.
"Everything it pretends to be: built up gradually, it leads one from a forest to a treenail, and
throws in, as a makeweight, a host of material concerning bricks, columns, cisterns, &c."—*English
Mechanic.*

DECORATIVE ARTS, etc.

Woods and Marbles (Imitation of).

SCHOOL OF PAINTING FOR THE IMITATION OF WOODS AND MARBLES, as Taught and Practised by A. R. Van der Burg and P. Van der Burg, Directors of the Rotterdam Painting Institution. Royal folio, 18¼ by 12¼ in., Illustrated with 24 full-size Coloured Plates; also 12 plain Plates, comprising 154 Figures. Second and Cheaper Edition. Price £1 11s. 6d.

List of Plates.

1. Various Tools required for Wood Painting —2, 3. Walnut: Preliminary Stages of Graining and Finished Specimen—4. Tools used for Marble Painting and Method of Manipulation—6. St. Remi Marble: Earlier Operations and Finished Specimen—7. Methods of Sketching different Grains, Knots, &c.—8, 9. Ash: Preliminary Stages and Finished Specimen — 10. Methods of Sketching Marble Grains—11, 12. Breche Marble: Preliminary Stages of Working and Finished Specimen—13. Maple: Methods of Producing the different Grains—14, 15. Bird's-eye Maple: Preliminary Stages and Finished Specimen—16. Methods of Sketching the different Species of White Marble—17, 18. White Marble: Preliminary Stages of Process and Finished Specimen—19. Mahogany: Specimens of various Grains and Methods of Manipulation —20, 21. Mahogany: Earlier Stages and Finished Specimen—22, 23, 24. Sienna Marble: Varieties of Grain, Preliminary Stages and Finished Specimen—25, 26, 27. Juniper Wood: Methods of producing Grain, &c.: Preliminary Stages and Finished Specimen—28, 29, 30. Vert de Mer Marble: Varieties of Grain and Methods of Working Unfinished and Finished Specimens—31, 32, 33. Oak: Varieties of Grain, Tools Employed, and Methods of Manipulation, Preliminary Stages and Finished Specimen—34, 35, 36. Waulsort Marble: Varieties of Grain, Unfinished and Finished Specimens.

"Those who desire to attain skill in the art of painting woods and marbles will find advantage consulting this book. . . . Some of the Working Men's Clubs should give their young men the opportunity to study it."—*Builder.*

"A comprehensive guide to the art. The explanations of the processes, the manipulation and management of the colours, and the beautifully executed plates will not be the least valuable to the student who aims at making his work a faithful transcript of nature."—*Building News.*

Wall Paper.

WALL PAPER DECORATION. By Arthur Seymour Jennings, Author of "Practical Paper Hanging." With numerous Illustrations. Demy 8vo [In preparation.

House Decoration.

ELEMENTARY DECORATION. A Guide to the Simpler Forms of Everyday Art. Together with PRACTICAL HOUSE DECORATION. By James W. Facey. With numerous Illustrations. In One Vol., 5s. strongly half-bound

House Painting, Graining, etc.

HOUSE PAINTING, GRAINING, MARBLING, AND SIGN WRITING, A Practical Manual of. By Ellis A. Davidson. Sixth Edition With Coloured Plates and Wood Engravings. 12mo, 6s. cloth boards.

"A mass of information, of use to the amateur and of value to the practical man."—*English Mechanic*

Decorators, Receipts for.

THE DECORATOR'S ASSISTANT : A Modern Guide to Decorative Artists and Amateurs, Painters, Writers, Gilders, &c. Containing upwards of 600 Receipts, Rules and Instructions ; with a variety of Information for General Work connected with every Class of Interior and Exterior Decorations, &c. Sixth Edition. 152 pp., crown 8vo, 1s. in wrapper

"Full of receipts of value to decorators, painters, gilders, &c. The book contains the gist of larger treatises on colour and technical processes. It would be difficult to meet with a work so full of varied information on the painter's art."—*Building News.*

Moyr Smith on Interior Decoration.

ORNAMENTAL INTERIORS, ANCIENT AND MODERN. By J. Moyr Smith. Super-royal 8vo, with 32 full-page Plates and numerous smaller Illustrations, handsomely bound in cloth, gilt top, price 18s.

"The book is well illustrated and handsomely got up, and contains some true criticism and good many good examples of decorative treatment."—*The Builder.*

British and Foreign Marbles.

MARBLE DECORATION *and the Terminology of British and Foreign Marbles.* A Handbook for Students. By GEORGE H. BLAGROVE, Author of " Shoring and its Application," &c. With 29 Illustrations. Crown 8vo, 3s. 6d. cloth.

" This most useful and much wanted handbook should be in the hands of every architect and builder."—*Building World.*

" A carefully and usefully written treatise ; the work is essentially practical."—*Scotsman*

Marble Working, etc.

MARBLE AND MARBLE WORKERS : A Handbook for Architects, Artists, Masons, and Students. By ARTHUR LEE, Author of " A Visit to Carrara," " The Working of Marble," &c. Small crown 8vo, 2s. cloth.

" A really valuable addition to the technical literature of architects and masons."—*Building News.*

DELAMOTTE'S WORKS ON ILLUMINATION AND ALPHABETS.

A PRIMER OF THE ART OF ILLUMINATION, *for the Use of Beginners:* with a Rudimentary Treatise on the Art, Practical Directions for its Exercise, and Examples taken from Illuminated MSS., printed in Gold and Colours. By F. DELAMOTTE. New and Cheaper Edition. Small 4to, 6s. ornamental boards.

" The examples of ancient MSS. recommended to the student, which, with much good sense, the author chooses from collections accessible to all, are selected with judgment and knowledge, as well as taste."—*Athenæum.*

ORNAMENTAL ALPHABETS, *Ancient and Mediæval, from the Eighth Century, with Numerals;* including Gothic, Church-Text, large and small, German, Italian, Arabesque, Initials for Illumination, Monograms, Crosses, &c. &c., for the use of Architectural and Engineering Draughtsmen Missal Painters, Masons, Decorative Painters, Lithographers, Engravers Carvers, &c. &c. Collected and Engraved by F. DELAMOTTE, and printed in Colours. New and Cheaper Edition. Royal 8vo, oblong, 2s. 6d. ornamental boards.

" For those who insert enamelled sentences round gilded chalices, who blazon shop legends over shop-doors, who letter church walls with pithy sentences from the Decalogue, this book will be useful."—*Athenæum.*

EXAMPLES OF MODERN ALPHABETS, *Plain and Ornamental;* including German, Old English, Saxon, Italic, Perspective, Greek, Hebrew, Court Hand, Engrossing, Tuscan, Riband, Gothic, Rustic, and Arabesque; with several Original Designs, and an Analysis of the Roman and Old English Alphabets, large and small, and Numerals, for the use of Draughtsmen, Surveyors, Masons, Decorative Painters, Lithographers, Engravers, Carvers, &c. Collected and Engraved by F. DELAMOTTE, and printed in Colours. New and Cheaper Edition. Royal 8vo, oblong, 2s. 6d. ornamental boards.

" There is comprised in it every possible shape into which the letters of the alphabet and numerals can be formed, and the talent which has been expended in the conception of the various plain and ornamental letters is wonderful."—*Standard.*

MEDIÆVAL ALPHABETS AND INITIALS FOR ILLUMINATORS. By F. G. DELAMOTTE. Containing 21 Plates and Illuminated Title, printed in Gold and Colours. With an Introduction by J. WILLIS BROOKS. Fourth and Cheaper Edition. Small 4to, 4s. ornamental boards.

" A volume in which the letters of the alphabet come forth glorified in gilding and all the colours of the prism interwoven and intertwined and intermingled."—*Sun.*

THE EMBROIDERER'S BOOK OF DESIGN. Containing Initials, Emblems, Cyphers, Monograms, Ornamental Borders, Ecclesiastical Devices, Mediæval and Modern Alphabets, and National Emblems. Collected by F. DELAMOTTE, and printed in Colours. Oblong royal 8vo, 1s. 6d. ornamental wrapper.

" The book will be of great assistance to ladies and young children who are endowed with the art of plying the needle in this most ornamental and useful pretty work."—*East Anglian Times.*

Wood Carving.

INSTRUCTIONS IN WOOD-CARVING, *for Amateurs:* with Hints on Design. By A LADY. With Ten Plates. New and Cheaper Edition. Crown 8vo, 2s. in emblematic wrapper.

" The handicraft of the wood-carver, so well as a book can impart it, may be learnt from ' A Lady's' publication."—*Athenæum.*

NATURAL SCIENCE, etc.

The Heavens and their Origin.

THE VISIBLE UNIVERSE: Chapters on the Origin and Construction of the Heavens. By J. E. GORE, F.R.A.S. Illustrated by 6 Stellar Photographs and 12 Plates. 8vo, 16s. cloth.

"A valuable and lucid summary of recent astronomical theory, rendered more valuable and attractive by a series of stellar photographs and other illustrations."—*The Times.*

"In presenting a clear and concise account of the present state of our knowledge, Mr. Gore has made a valuable addition to the literature of the subject."—*Nature.*

"As interesting as a novel, and instructive withal; the text being made still more luminous by stellar photographs and other illustrations. . . . A most valuable book."—*Manchester Examiner.*

"One of the finest works on astronomical science that has recently appeared in our language.
Leeds Mercury.

The Constellations.

STAR GROUPS: A Student's Guide to the Constellations. By J. ELLARD GORE, F.R.A.S., M.R.I.A., &c., Author of "The Visible Universe," "The Scenery of the Heavens." With 30 Maps. Small 4to, 5s. cloth, silvered.

"A knowledge of the principal constellations visible in our latitudes may be easily acquired from the thirty maps and accompanying text contained in this work."—*Nature*

"The volume contains thirty maps showing stars of the sixth magnitude—the usual naked-eye limit—and each is accompanied by a brief commentary, adapted to facilitate recognition and bring to notice objects of special interest. For the purpose of a preliminary survey of the ' midnight pomp ' of the heavens, nothing could be better than a set of delineations averaging scarcely twenty square inches in area, and including nothing that cannot at once be identified."—*Saturday Review.*

"A very compact and handy guide to the constellations."—*Athenæum.*

Astronomical Terms.

AN ASTRONOMICAL GLOSSARY; or, Dictionary of Terms used in Astronomy. With Tables of Data and Lists of Remarkable and Interesting Celestial Objects. By J. ELLARD GORE, F.R.A.S., Author of "The Visible Universe," &c. Small crown 8vo, 2s. 6d. cloth.

"A very useful little work for beginners in astronomy, and not to be despised by more adj vanced students."—*The Times.*

"Astronomers of all kinds will be glad to have it for reference."—*Guardian.*

The Microscope.

THE MICROSCOPE: Its Construction and Management, including Technique, Photo-micrography, and the Past and Future of the Microscope. By Dr. HENRI VAN HEURCK. Re-Edited and Augmented from the Fourth French Edition, and Translated by WYNNE E. BAXTER, F.G.S. 400 pages, with upwards of 250 Woodcuts. Imp. 8vo, 18s. cloth.

"A translation of a well-known work, at once popular and comprehensive."—*Times.*

"The translation is as felicitous as it is accurate."—*Nature.*

The Microscope.

PHOTO-MICROGRAPHY. By Dr. H. VAN HEURCK. Extracted from the above Work. Royal 8vo, with Illustrations, 1s. sewed.

Astronomy.

ASTRONOMY. By the late Rev. ROBERT MAIN, F.R.S. Third Edition, Revised, by WM. T. LYNN, B.A., F.R.A.S., 12mo, 2s. cloth.

"A sound and simple treatise, and a capital book for beginners."—*Knowledge.*

Recent and Fossil Shells.

A MANUAL OF THE MOLLUSCA: Being a Treatise on Recent and Fossil Shells. By S. P. WOODWARD, A.L.S., F.G.S. With an Appendix on Recent and Fossil Conchological Discoveries, by RALPH TATE, A.L.S., F.G.S. With 23 Plates and upwards of 300 Woodcuts. Reprint of Fourth Edition, 1880. Crown 8vo, 7s. 6d. cloth.

"A most valuable storehouse of conchological and geological information."—*Science Gossip.*

Geology and Genesis.

THE TWIN RECORDS OF CREATION; or, Geology and Genesis: their Perfect Harmony and Wonderful Concord. By GEORGE W. VICTOR LE VAUX. Fcap. 8vo, 5s. cloth.

"A valuable contribution to the evidences of Revelation, and disposes very conclusively of the arguments of those who would set God's Works against God's Word. No real difficulty is shirked and no sophistry is left unexposed."—*The Rock.*

Geology.

RUDIMENTARY TREATISE ON GEOLOGY, PHYSICAL AND HISTORICAL. With especial reference to the British series of Rocks. By R. TATE, F.G.S. With 250 Illustrations. 12mo, 5s. cloth boards

DR. LARDNER'S COURSE OF NATURAL PHILOSOPHY.

HANDBOOK OF MECHANICS. Re-written and Enlarged by
B. Loewy, F.R.A.S. Post 8vo, 6s, cloth.

"Mr. Loewy has carefully revised the book, and brought it up to modern requirements."—
Nature.

HANDBOOK OF HYDROSTATICS & PNEUMATICS. Enlarged
by B. Loewy, F.R.A.S. Post 8vo, 5s. cloth.

"For those 'who desire to attain an accurate knowledge of physical science without the profound methods of mathematical investigation,' this work is well adapted."—*Chemical News.*

HANDBOOK OF HEAT. Edited and almost entirely Re-written
by Benjamin Loewy, F.R.A.S., &c. Post 8vo, 6s. cloth.

"The style is always clear and precise, and conveys instruction without leaving any cloudiness or lurking doubts behind."—*Engineering.*

HANDBOOK OF OPTICS. By Dr. Lardner. Edited by T. O.
Harding, B.A. Post 8vo, 5s. cloth.

"Written by an able scientific writer and beautifully illustrated."—*Mechanic's Magazine.*

HANDBOOK OF ELECTRICITY AND MAGNETISM. By Dr.
Lardner. Edited by G. C. Foster, B.A. Post 8vo, 5s. cloth.

"The book could not have been entrusted to anyone better calculated to preserve the terse and lucid style of Lardner."—*Popular Science Review.*

HANDBOOK OF ASTRONOMY. By Dr. Lardner. Fourth
Edition by E. Dunkin, F.R.A.S. Post 8vo, 9s. 6d. cloth.

"Probably no other book contains the same amount of information in so compendious and well-arranged a form—certainly none at the price at which this is offered to the public."—*Athenæum.*

"We can do no other than pronounce this work a most valuable manual of astronomy, and we strongly recommend it to all who wish to acquire a general—but at the same time correct—acquaintance with this sublime science."—*Quarterly Journal of Science.*

DR. LARDNER'S MUSEUM OF SCIENCE AND ART.

THE MUSEUM OF SCIENCE AND ART. Edited by
Dr. Lardner. With upwards of 1,200 Engravings on Wood. In 6 Double
Volumes, £1 1s. in a new and elegant cloth binding ; or handsomely bound in
half-morocco, 31s. 6d.

"A cheap and interesting publication, alike informing and attractive. The papers combine subjects of importance and great scientific knowledge, considerable inductive powers, and a popular style of treatment."—*Spectator.*

The 'Museum of Science and Art' is the most valuable contribution that has ever been made to the Scientific Instruction of every class of society."—Sir David Brewster, in the *North British Review.*

** *Separate books formed from the above, fully Illustrated, suitable for
Workmen's Libraries, Science Classes, etc.*

Common Things Explained. 5s.	**Steam and its Uses.** 2s. cloth.
The Microscope. 2s. cloth.	**Popular Astronomy.** 4s. 6d. cloth.
Popular Geology. 2s. 6d. cloth.	**The Bee and White Ants.** 2s. cloth.
Popular Physics. 2s. 6d. cloth.	**The Electric Telegraph.** 1s.

Dr. Lardner's School Handbooks.

NATURAL PHILOSOPHY FOR SCHOOLS. Fcap. 8vo, 3s. 6d

"A very convenient class-book for junior students in private schools."—*British Quarterly Review.*

ANIMAL PHYSIOLOGY FOR SCHOOLS. Fcap. 8vo, 3s. 6d.

"Clearly written, well arranged, and excellently illustrated."—*Gardener's Chronicle.*

THE ELECTRIC TELEGRAPH. By Dr. Lardner. Revised by E. B. Bright, F.R.A.S. Fcap. 8vo, 2s. 6d. cloth.

"One of the most readable books extant on the Electric Telegraph."—*English Mechanic.*

D

CHEMICAL MANUFACTURES, CHEMISTRY.

Chemistry for Engineers, etc.

ENGINEERING CHEMISTRY: A Practical Treatise for the Use of Analytical Chemists, Engineers, Iron Masters, Iron Founders, Students, and others. Comprising Methods of Analysis and Valuation of the Principal Materials used in Engineering Work, with numerous Analyses, Examples, and Suggestions. By H. JOSHUA PHILLIPS, F.I.C., F.C.S. formerly Analytical and Consulting Chemist to the Great Eastern Railway. Second Edition, Revised and Enlarged. Crown 8vo, 400 pp., with Illustrations, 10s. 6d. cloth.   [*Just published.*

"In this work the author has rendered no small service to a numerous body of practical men. . . . The analytical methods may be pronounced most satisfactory, being as accurate as the despatch required of engineering chemists permits."—*Chemical News.*

"The book will be very useful to those who require a handy and concise *resume* of approved methods of analysing and valuing metals, oils, fuels, &c. It is, in fact, a work for chemists, a guide to the routine of the engineering laboratory. . . . The book is full of good things. As a hand-book of technical analysis, it is very welcome."—*Builder.*

"The analytical methods given are, as a whole, such as are likely to give rapid and trustworthy results in experienced hands. There is much excellent descriptive matter in the work, the chapter on 'Oils and Lubrication' being specially noticeable in this respect."—*Engineer.*

Explosives and Dangerous Goods.

DANGEROUS GOODS: Their Sources and Properties, Modes of Storage, and Transport. With Notes and Comments on Accidents arising therefrom, together with the Government and Railway Classifications, Acts of Parliament, &c. A Guide for the use of Government and Railway Officials, Steamship Owners, Insurance Companies and Manufacturers and users of Explosives and Dangerous Goods. By H. JOSHUA PHILLIPS, F.I.C., F.C.S., Author of "Engineering Chemistry, &c." Crown 8vo, 350 pp., 9s. cloth.   [*Just ready.*

Alkali Trade, Manufacture of Sulphuric Acid, etc.

A MANUAL OF THE ALKALI TRADE, including the Manufacture of Sulphuric Acid, Sulphate of Soda, and Bleaching Powder. By JOHN LOMAS, Alkali Manufacturer, Newcastle-upon-Tyne and London. With 232 Illustrations and Working Drawings, and containing 390 pages of Text. Second Edition, with Additions. Super-royal 8vo, £1 10s. cloth.

"This book is written by a manufacturer for manufacturers. The working details of the most approved forms of apparatus are given, and these are accompanied by no less than 232 wood engravings, all of which may be used for the purposes of construction. Every step in the manufacture is very fully described in this manual, and each improvement explained."—*Athenæum.*

The Blowpipe.

THE BLOWPIPE IN CHEMISTRY, MINERALOGY, AND GEOLOGY. Containing all known Methods of Anhydrous Analysis, many Working Examples, and Instructions for Making Apparatus. By Lieut.-Colonel W. A. Ross, R.A., F.G.S. With 120 Illustrations. Second Edition, Revised and Enlarged. Crown 8vo, 5s. cloth.

"The student who goes through the course of experimentation here laid down will gain a better insight into inorganic chemistry and mineralogy than if he had 'got up' any of the best text-books, and passed any number of examinations in their contents."—*Chemical News.*

Commercial Chemical Analysis.

THE COMMERCIAL HANDBOOK OF CHEMICAL ANALYSIS; or, Practical Instructions for the determination of the Intrinsic or Commercial Value of Substances used in Manufactures, in Trades, and in the Arts. By A. NORMANDY. New Edition, by H.M. NOAD, F.R.S. Crown 8vo, 12s. 6d. cloth.

"We strongly recommend this book to our readers as a guide, alike indispensable to the ousewife as to the pharmaceutical practitioner."—*Medical Times.*

Dye-Wares and Colours.

THE MANUAL OF COLOURS AND DYE-WARES: Their Properties, Applications, Valuations, Impurities, and Sophistications. For the use of Dyers, Printers, Drysalters, Brokers, &c. By J. W. SLATER. Second Edition, Revised and greatly Enlarged. Crown 8vo, 7s. 6d. cloth.

"A complete encyclopaedia of the *materia tinctoria.* The information given respecting each article is full and precise, and the methods of determining their value are given with clearness, and are practical as well as valuable."—*Chemist and Druggist.*

Modern Brewing and Malting.

A HANDYBOOK FOR BREWERS: Being a Practical Guide to the Art of Brewing and Malting. Embracing the Conclusions of Modern Research which bear upon the Practice of Brewing. By HERBERT EDWARDS WRIGHT, M.A. Crown 8vo, 530 pp., 12s. 6d. cloth.

"May be consulted with advantage by the student who is preparing himself for examinational tests, while the scientific brewer will find in it a *resume* of all the most important discoveries of modern times. The work is written throughout in a clear and concise manner, and the author takes great care to discriminate between vague theories and well-established facts."—*Brewers' Journal.*

"We have great pleasure in recommending this handybook, and have no hesitation in saying that it is one of the best—if not the best—which has yet been written on the subject of beer-brewing in this country, and it should have a place on the shelves of every brewer's library."—*The Brewer's Guardian.*

"Although the requirements of the student are primarily considered, an acquaintance of half-an-hour's duration cannot fail to impress the practical brewer with the sense of having found a trustworthy guide and practical counsellor in brewery matters."—*Chemical Trade Journal.*

Analysis and Valuation of Fuels.

FUELS: SOLID, LIQUID, AND GASEOUS, Their Analysis and Valuation. For the Use of Chemists and Engineers. By H. J. PHILLIPS, F.C.S., formerly Analytical and Consulting Chemist to the Great Eastern Railway. Second Edition, Revised and Enlarged. Crown 8vo, 5s. cloth.

"Ought to have its place in the laboratory of every metallurgical establishment, and wherever fuel is used on a large scale."—*Chemical News.*

"Cannot fail to be of wide interest, especially at the present time."—*Railway News.*

Pigments.

THE ARTIST'S MANUAL OF PIGMENTS. Showing their Composition, Conditions of Permanency, Non-Permanency, and Adulterations; Effects in Combination with Each Other and with Vehicles; and the most Reliable Tests of Purity Together with the Science and Art Department's Examination Questions on Painting. By H. C. STANDAGE. Second Edition. Crown 8vo, 2s. 6d. cloth.

"This work is indeed *multum-in-parvo*, and we can, with good conscience, recommend it to all who come in contact with pigments, whether as makers, dealers or users."—*Chemical Review.*

Gauging. Tables and Rules for Revenue Officers, Brewers, etc.

A POCKET BOOK OF MENSURATION AND GAUGING: Containing Tables, Rules and Memoranda for Revenue Officers, Brewers, Spirit Merchants, &c. By J. B. MANT (Inland Revenue). Second Edition, Revised. 18mo, 4s. cloth.

"This handy and useful book is adapted to the requirements of the Inland Revenue Department, and will be a favourite book of reference."—*Civilian.*

"Should be in the hands of every practical brewer."—*Brewers' Journal.*

INDUSTRIAL ARTS, TRADES, AND MANUFACTURES.

Cotton Spinning.

COTTON MANUFACTURE: A Manual of Practical Instruction in the Processes of Opening, Carding, Combing, Drawing, Doubling and Spinning of Cotton, the Methods of Dyeing, &c. For the Use of Operatives, Overlookers and Manufacturers. By JOHN LISTER, Technical Instructor, Pendleton. 8vo, 7s. 6d. cloth. [*Just published.*

"This invaluable volume is a distinct advance in the literature of cotton manufacture."—*Machinery.*

"It is thoroughly reliable, fulfilling nearly all the requirements desired."—*Glasgow Herald.*

Flour Manufacture, Milling, etc.

FLOUR MANUFACTURE: A Treatise on Milling Science and Practice. By FRIEDRICH KICK, Imperial Regierungsrath, Professor of Mechanical Technology in the Imperial German Polytechnic Institute, Prague. Translated from the Second Enlarged and Revised Edition with Supplement. By H. H. P. POWLES, A.-M.Inst.C.E. Nearly 400 pp. Illustrated with 28 Folding Plates, and 167 Woodcuts. Royal 8vo, 25s. cloth.

"This valuable work is, and will remain, the standard authority on the science of milling. . . The miller who has read and digested this work will have laid the foundation, so to speak, of a successful career; he will have acquired a number of general principles which he can proceed to apply. In this handsome volume we at last have the accepted text-book of modern milling in good, sound English, which has little, if any, trace of the German idiom."—*The Miller.*

"The appearance of this celebrated work in English is very opportune, and British millers will, we are sure, not be slow in availing themselves of its pages."—*Millers' Gazette.*

Agglutinants.

CEMENTS, PASTES, GLUES AND GUMS: A Practical Guide to the Manufacture and Application of the various Agglutinants required in the Building, Metal-Working, Wood-Working and Leather-Working Trades, and for Workshop, Laboratory or Office Use. With upwards of 900 Recipes and Formulæ. By H. C. STANDAGE, Chemist. Crown 8vo, 2s. 6d. cloth. [*Just published.*

"We have pleasure in speaking favourably of this volume. So far as we have had experience, which is not inconsiderable, this manual is trustworthy."—*Athenæum.*

"As a revelation of what are considered trade secrets, this book will arouse an amount of curiosity among the large number of industries it touches."—*Daily Chronicle.*

"In this goodly collection of receipts it would be strange if a cement for any purpose cannot be found."—*Oil and Colourman's Journal.*

Soap-making.

THE ART OF SOAP-MAKING: A Practical Handbook of the Manufacture of Hard and Soft Soaps, Toilet Soaps, etc. Including many New Processes, and a Chapter on the Recovery of Glycerine from Waste Leys. By ALEXANDER WATT. Fourth Edition, Enlarged. Crown 8vo, 7s. 6d. cloth.

"The work will prove very useful, not merely to the technological student, but to the practical soap-boiler who wishes to understand the theory of his art."—*Chemical News.*

"A thoroughly practical treatise on an art which has almost no literature in our language. We congratulate the author on the success of his endeavour to fill a void in English technical literature."—*Nature.*

Paper Making.

PRACTICAL PAPER-MAKING: A Manual for Paper-makers and Owners and Managers of Paper-Mills. With Tables, Calculations, &c. By G. CLAPPERTON, Paper-maker. With Illustrations of Fibres from Micro-Photographs. Crown 8vo, 5s. cloth. [*Just published.*

"The author caters for the requirements of responsible mill hands, apprentices, &c., whilst his manual will be found of great service to students of technology, as well as to veteran paper makers and mill owners. The illustrations form an excellent feature."—*Paper Trade Review.*

"We recommend everybody interested in the trade to get a copy of this thoroughly practical book."—*Paper Making.*

Paper Making.

THE ART OF PAPER MAKING: A Practical Handbook of the Manufacture of Paper from Rags, Esparto, Straw, and other Fibrous Materials, Including the Manufacture of Pulp from Wood Fibre, with a Description of the Machinery and Appliances used. To which are added Details of Processes for Recovering Soda from Waste Liquors. By ALEXANDER WATT, Author of "The Art of Soap-Making" With Illusts. Crown 8vo, 7s. 6d. cloth.

"It may be regarded as the standard work on the subject. The book is full of valuable information. The 'Art of Paper-making,' is in every respect a model of a text-book, either for a technical class or for the private student."—*Paper and Printing Trades Journal.*

Leather Manufacture.

THE ART OF LEATHER MANUFACTURE. Being a Practical Handbook, in which the Operations of Tanning, Currying, and Leather Dressing are fully Described, and the Principles of Tanning Explained, and many Recent Processes Introduced; as also the Methods for the Estimation of Tannin, and a Description of the Arts of Glue Boiling, Gut Dressing, &c. By ALEXANDER WATT, Author of "Soap-Making," &c. Second Edition. Crown 8vo, 9s. cloth.

"A sound, comprehensive treatise on tanning and its accessories. It is an eminently valuable production, which redounds to the credit of both author and publishers."—*Chemical Review.*

Boot and Shoe Making.

THE ART OF BOOT AND SHOE-MAKING. A Practical Handbook, including Measurement, Last-Fitting, Cutting-Out, Closing, and Making, with a Description of the most approved Machinery employed. By JOHN B. LENO, late Editor of *St. Crispin*, and *The Boot and Shoe-Maker.* 12mo, 2s. cloth limp.

"This excellent treatise is by far the best work ever written. The chapter on clicking, which shows how waste may be prevented, will save fifty times the price of the book." *Scottish Leather Trader.*

Wood Engraving.

WOOD ENGRAVING: A Practical and Easy Introduction to the Study of the Art. By WILLIAM NORMAN BROWN. Second Edition. With numerous Illustrations. 12mo, 1s. 6d. cloth limp.

"The book is clear and complete, and will be useful to anyone wanting to understand the first elements of the beautiful art of wood engraving."—*Graphic.*

Watch Adjusting.

THE WATCH ADJUSTER'S MANUAL: A Practical Guide for the Watch and Chronometer Adjuster in Making, Springing, Timing and Adjusting for Isochronism, Positions and Temperatures. By C. E. FRITTS. 370 pages, with Illustrations, 8vo, 16s. cloth. *[Just published.*

Horology.

A TREATISE ON MODERN HOROLOGY, in Theory and Practice. Translated from the French of CLAUDIUS SAUNIER, ex-Director of the School of Horology at Maçon, by JULIEN TRIPPLIN, F.R.A.S., Besançon Watch Manufacturer, and EDWARD RIGG, M.A., Assayer in the Royal Mint. With 78 Woodcuts and 22 Coloured Copper Plates. Second Edition. Super-royal 8vo, £2 2s. cloth ; £2 10s. half-calf.

"There is no horological work in the English language at all to be compared to this production of M. Saunier's for clearness and completeness. It is alike good as a guide for the student and as a reference for the experienced horologist and skilled workman."—*Horological Journal.*

"The latest, most complete, and the most reliable of those literary productions to which continental watchmakers are indebted for the mechanical superiority over their English brethren—in fact, the Book of Books, is M. Saunier's 'Treatise.'"—*Watchmaker, Jeweller and Silversmith.*

Watchmaking.

THE WATCHMAKER'S HANDBOOK. Intended as a Workshop Companion for those engaged in Watchmaking and the Allied Mechanical Arts. Translated from the French of CLAUDIUS SAUNIER, and considerably enlarged by JULIEN TRIPPLIN, F.R.A.S., Vice-President of the Horological Institute, and EDWARD RIGG, M.A., Assayer in the Royal Mint. With numerous Woodcuts and 14 Copper Plates. Third Edition. Crown 8vo, 9s. cloth.

"Each part is truly a treatise in itself. The arrangement is good and the language is clear and concise. It is an admirable guide for the young watchmaker."—*Engineering.*

"It is impossible to speak too highly of its excellence. It fulfils every requirement in a handbook intended for the use of a workman. Should be found in every workshop."—*Watch and Clockmaker.*

"This book contains an immense number of practical details bearing on the daily occupation of a watchmaker."—*Watchmaker and Metalworker* (Chicago).

Watches and Timekeepers.

A HISTORY OF WATCHES AND OTHER TIMEKEEPERS. By JAMES F. KENDAL, M.B.H.Inst. 1s. 6d. boards; or 2s. 6d. cloth gilt.

"Mr. Kendal's book, for its size, is the best which has yet appeared on this subject in the English language."—*Industries.*

"Open the book where you may, there is interesting matter in it concerning the ingenious devices of the ancient or modern horologer. The subject is treated in a liberal and entertaining spirit, as might be expected of a historian who is a master of the craft."—*Saturday Review.*

Electrolysis of Gold, Silver, Copper, etc.

ELECTRO-DEPOSITION : A Practical Treatise on the Electrolysis of Gold, Silver, Copper, Nickel, and other Metals and Alloys. With descriptions of Voltaic Batteries, Magneto and Dynamo-Electric Machines, Thermopiles, and of the Materials and Processes used in every Department of the Art, and several Chapters on Electro-Metallurgy. By ALEXANDER WATT, Author of "Electro-Metallurgy," &c. Third Edition, Revised. Crown 8vo, 9s. cloth.

"Eminently a book for the practical worker in electro-deposition. It contains practical descriptions of methods, processes and materials as actually pursued and used in the workshop."—*Engineer.*

Electro-Metallurgy.

ELECTRO-METALLURGY ; Practically Treated. By ALEXANDER WATT, Author of "Electro-Deposition," &c. Tenth Edition, including the most recent Processes. 12mo, 4s. cloth boards.

"From this book both amateur and artisan may learn everything necessary for the successful prosecution of electroplating."—*Iron.*

Working in Gold.

THE JEWELLER'S ASSISTANT IN THE ART OF WORKING IN GOLD: A Practical Treatise for Masters and Workmen, Compiled from the Experience of Thirty Years' Workshop Practice. By GEORGE E. GEE, Author of "The Goldsmith's Handbook," &c. Cr. 8vo, 7s. 6d. cloth.

"This manual of technical education is apparently destined to be a valuable auxiliary to a handicraft which is certainly capable of great improvement."—*The Times.*

"Very useful in the workshop, as the knowledge is practical, having been acquired by long experience, and all the recipes and directions are guaranteed to be successful."—*Jeweller and Metalworker.*

Electroplating.

ELECTROPLATING: A Practical Handbook on the Deposition of Copper, Silver, Nickel, Gold, Aluminium, Brass, Platinum, &c. &c. With Descriptions of the Chemicals, Materials, Batteries, and Dynamo Machines used in the Art. By J. W. URQUHART, C.E., Author of " Electric Light," &c. Third Edition, with Additions. Crown 8vo. 5s. cloth.
" An **excellent** practical manual."—*Engineering.*
" An excellent work, giving the newest information."—*Horological Journal.*

Electrotyping.

ELECTROTYPING: The Reproduction and Multiplication of Printing Surfaces and Works of Art by the Electro-deposition of Metals. By J. W. URQUHART, C.E. Crown 8vo, 5s. cloth.
" The book is thoroughly practical. The reader is, therefore, conducted through the leading laws of electricity, then through the metals used by electrotypers, the apparatus, and the depositing processes, up to the final preparation of the work."—*Art Journal.*

Goldsmiths' Work.

THE GOLDSMITH'S HANDBOOK. By GEORGE E. GEE, Jeweller, &c. Third Edition, considerably Enlarged. 12mo, 3s. 6d. cl. bds.
" A good, sound educator, which will be accepted as an authority."—*Horological Journal.*

Silversmiths' Work.

THE SILVERSMITH'S HANDBOOK. By GEORGE E. GEE, Jeweller, &c. Second Edition, Revised. 12mo, 3s. 6d. cloth.
" The chief merit of the work is its practical character. . . The workers in the trade will speedily discover its merits when they sit down to study it."—*English Mechanic.*
.* The above two works together, strongly half-bound, price 7s.

Sheet Metal Working.

THE SHEET METAL WORKER'S INSTRUCTOR: For Zinc, Sheet Iron, Copper, and Tin Plate Workers. Containing Rules for describing the Patterns required in the Different Branches of the Trade. By R. H. WARN, Tin Plate Worker. With Thirty-two Plates. 8vo, 7s. 6d. cl.

Bread and Biscuit Baking.

THE BREAD AND BISCUIT BAKER'S AND SUGAR-BOILER'S ASSISTANT. Including a large variety of Modern Recipes. By ROBERT WELLS, Practical Baker. Crown 8vo, 2s. cloth.
" A large number of wrinkles for the ordinary cook, as well as the baker."—*Saturday Review.*

Confectionery for Hotels and Restaurants.

THE PASTRYCOOK AND CONFECTIONER'S GUIDE. For Hotels, Restaurants and the Trade in general, adapted also for Family Use. By ROBERT WELLS. Crown 8vo, 2s. cloth.
" We cannot speak too highly of this really excellent work. In these days of keen competition our readers cannot do better than purchase this book."—*Bakers' Times.*

Ornamental Confectionery.

ORNAMENTAL CONFECTIONERY: A Guide for Bakers, Confectioners and Pastrycooks; including a variety of Modern Recipes, and Remarks on Decorative and Coloured Work. With 129 Original Designs. By ROBERT WELLS. Crown 8vo, cloth gilt, 5s.
" A valuable work, practical, and should be in the hands of every baker and confectioner. The illustrative designs are alone worth treble the amount charged for the whole work."—*Bakers' Times.*

Flour Confectionery.

THE MODERN FLOUR CONFECTIONER. Wholesale and Retail. Containing a large Collection of Recipes for Cheap Cakes, Biscuits, &c. With Remarks on the Ingredients used in their Manufacture. By R. WELLS. Crown 8vo, 2s. cloth.
" The work is of a decidedly practical character, and in every recipe regard is had to economical working."—*North British Daily Mail.*

Laundry Work.

LAUNDRY MANAGEMENT. A Handbook for Use in Private and Public Laundries, Including Descriptive Accounts of Modern Machinery and Appliances for Laundry Work. By the EDITOR of " The Laundry Journal." Second Edition. Crown 8vo, 2s. 6d. cloth.
" This book should certainly occupy an honoured place on the shelves of all housekeepers who wish to keep themselves *au courant* of the newest appliances and methods."—*The Queen.*

HANDYBOOKS FOR HANDICRAFTS.

By PAUL N. HASLUCK,

EDITOR OF "WORK" (NEW SERIES); AUTHOR OF "LATHEWORK," "MILLING MACHINES," &c.

Crown 8vo, 144 pages, cloth, price 1s. each.

☞ *These* HANDYBOOKS *have been written to supply information for* WORKMEN, STUDENTS, *and* AMATEURS *in the several Handicrafts, on the actual* PRACTICE *of the* WORKSHOP, *and are intended to convey in plain language* TECHNICAL KNOW-LEDGE *of the several* CRAFTS. *In describing the processes employed, and the manipulation of material, workshop terms are used; workshop practice is fully explained; and the text is freely illustrated with drawings of modern tools, appliances, and processes.*

THE METAL TURNER'S HANDYBOOK. A Practical Manual *for Workers at the Foot-Lathe.* With over 100 Illustrations. Price 1s.
"The book will be of service alike to the amateur and the artisan turner. It displays thorough knowledge of the subject."—*Scotsman.*

THE WOOD TURNER'S HANDYBOOK. A Practical Manual *for Workers at the Lathe.* With over 100 Illustrations. Price 1s.
"We recommend the book to young turners and amateurs. A multitude of workmen have hitherto sought in vain for a manual of this special industry."—*Mechanical World.*

THE WATCH JOBBER'S HANDYBOOK. A Practical Manual *on Cleaning, Repairing, and Adjusting.* With upwards of 100 Illustrations. Price 1s.
"We strongly advise all young persons connected with the watch trade to acquire and study this inexpensive work."—*Clerkenwell Chronicle.*

THE PATTERN MAKER'S HANDYBOOK. A Practical Manual on the Construction of Patterns for Founders. With upwards of 100 Illustrations. Price 1s.
"A most valuable, if not indispensable, manual for the pattern maker."—*Knowledge.*

THE MECHANIC'S WORKSHOP HANDYBOOK. A Practical Manual *on Mechanical Manipulation.* Embracing Information on various Handicraft Processes, with Useful Notes and Miscellaneous Memoranda. Comprising about 200 Subjects. Price 1s.
"A very clever and useful book, which should be found in every workshop; and it should certainly find a place in all technical schools."—*Saturday Review.*

THE MODEL ENGINEER'S HANDYBOOK. A Practical Manual on the Construction of Model Steam Engines. With upwards of 100 Illustrations. Price 1s.
"Mr. Hasluck has produced a very good little book."—*Builder.*

THE CLOCK JOBBER'S HANDYBOOK. A Practical Manual *on Cleaning, Repairing, and Adjusting.* With upwards of 100 Illustrations. Price 1s.
"It is of inestimable service to those commencing the trade."—*Coventry Standard.*

THE CABINET WORKER'S HANDYBOOK : A Practical Manual on the Tools, Materials, Appliances, and Processes employed in Cabinet Work. With upwards of 100 Illustrations. Price 1s.
"Mr. Hasluck's thoroughgoing little Handybook is amongst the most practical guides we have seen for beginners in cabinet-work."—*Saturday Review.*

THE WOODWORKER'S HANDYBOOK OF MANUAL IN-STRUCTION. Embracing Information on the Tools, Materials, Appliances and Processes employed in Woodworking. With 104 Illustrations. Price 1s
[*Just published.*

THE METALWORKER'S HANDYBOOK. With upwards of 100 Illustrations. [*In preparation.*

***⁎⁎* OPINIONS OF THE PRESS.**

"Written by a man who knows, not only how work ought to be done, but how to do it, and how to convey his knowledge to others."—*Engineering.*
"Mr. Hasluck writes admirably, and gives complete instructions."—*Engineer.*
"Mr. Hasluck combines the experience of a practical teacher with the manipulative skill and scientific knowledge of processes of the trained mechanician, and the manuals are marvels of what can be produced at a popular price."—*Schoolmaster.*
"Helpful to workmen of all ages and degrees of experience."—*Daily Chronicle.*
"Practical, sensible, and remarkably cheap."—*Journal of Education.*
"Concise, clear and practical."—*Saturday Review.*

COMMERCE, COUNTING-HOUSE WORK, TABLES, etc.

Commercial French.
A NEW BOOK OF COMMERCIAL FRENCH : Grammar—Vocabulary — Correspondence — Commercial Documents — Geography — Arithmetic—Lexicon. By P. CARROUÉ, Professor in the City High School J.—B. Say (Paris). Crown 8vo, 4s. 6d. cloth.   [*Just published*

Commercial Education.
LESSONS IN COMMERCE. By Professor R. GAMBARO, of the Royal High Commercial School at Genoa. Edited and Revised by JAMES GAULT, Professor of Commerce and Commercial Law in King's College, London. Second Edition, Revised. Crown 8vo, 3s. 6d. cloth.[*Just published.*

"The publishers of this work have rendered considerable service to the cause of commercial education by the opportune production of this volume. . . . The work is peculiarly acceptable to English readers and an admirable addition to existing class-books. In a phrase, we think the work attains its object in furnishing a brief account of those laws and customs of British trade with which the commercial man interested therein should be familiar."—*Chamber of Commerce Journal.*
"An invaluable guide in the hands of those who are preparing for a commercial career."
Counting House.

Foreign Commercial Correspondence.
THE FOREIGN COMMERCIAL CORRESPONDENT : Being Aids to Commercial Correspondence in Five Languages—English, French, German, Italian, and Spanish. BY CONRAD E. BAKER. Second Edition. Crown 8vo, 3s. 6d. cloth.

"Whoever wishes to correspond in all the languages mentioned by Mr. Baker cannot do better than study this work, the materials of which are excellent and conveniently arranged. They consist not of entire specimen letters but—what are far more useful—short passages, sentences, or phrases expressing the same general idea in various forms."—*Athenæum.*
"A careful examination has convinced us that it is unusually complete, well arranged, and reliable. The book is a thoroughly good one."—*Schoolmaster.*

Accounts for Manufacturers.
FACTORY ACCOUNTS : Their Principles and Practice. A Handbook for Accountants and Manufacturers, with Appendices on the Nomenclature of Machine Details; the Income Tax Acts; the Rating of Factories; Fire and Boiler Insurance; the Factory and Workshop Acts, &c., including also a Glossary of Terms and a large number of Specimen Rulings. By EMILE GARCKE and J. M. FELLS. Fourth Edition, Revised and Enlarged. Demy 8vo, 250 pages, 6s. strongly bound.

"A very interesting description of the requirements of Factory Accounts. . . . the principle of assimilating the Factory Accounts to the general commercial books is one which we thoroughly agree with."—*Accountants' Journal.*
"Characterised by extreme thoroughness. There are few owners of factories who would not derive great benefit from the perusal of this most admirable work."—*Local Government Chronicle.*

Modern Metrical Units and Systems.
MODERN METROLOGY : A Manual of the Metrical Units and Systems of the Present Century. With an Appendix containing a proposed English System. By LOWIS D'A. JACKSON, A.M.Inst.C.E., Author of "Aid to Survey Practice," &c. Large crown 8vo, 12s. 6d. cloth.
"We recommend the work to all interested in the practical reform of our weights and measures."—*Nature.*

The Metric System and the British Standards.
A SERIES OF METRIC TABLES, in which the British Standard Measures and Weights are compared with those of the Metric System at present in Use on the Continent. By C. H. DOWLING, C.E. 8vo, 10s. 6d. strongly bound
"Mr. Dowling's Tables are well put together as a ready-reckoner for the conversion of one system into the other."—*Athenæum.*

Iron Shipbuilders' and Merchants' Weight Tables.
IRON-PLATE WEIGHT TABLES: For Iron Shipbuilders, Engineers, and Iron Merchants. Containing the Calculated Weights of upwards of 150,000 different sizes of Iron Plates, from 1 foot by 6 in. by ¼ in. to 10 feet by 5 feet by 1 in. Worked out on the basis of 40 lbs. to the square foot of Iron of 1 inch in thickness. Carefully compiled and thoroughly Revised by H. BURLINSON and W. H. SIMPSON. Oblong 4to, 25s. half-bound.
"This work will be found of great utility. The authors have had much practical experience of what is wanting in making estimates; and the use of the book will save much time in making elaborate calculations."—*English Mechanic.*

Chadwick's Calculator for Numbers and Weights Combined.

THE NUMBER, WEIGHT, AND FRACTIONAL CALCU-
LATOR. Containing upwards of 250,000 Separate Calculations, showing at
a glance the value at 422 different rates, ranging from $\frac{1}{16}$th of a Penny to
20s. each, or per cwt., and £20 per ton, of any number of articles consecu-
tively, from 1 to 470.—Any number of cwts., qrs., and lbs., from 1 cwt. to 470
cwts.—Any number of tons, cwts., qrs., and lbs., from 1 to 1,000 tons. By
WILLIAM CHADWICK, Public Accountant. Third Edition, Revised and Improved. 8vo,18s., strongly bound for Office wear and tear.

☞ *Is adapted for the use of Accountants and Auditors, Railway Companies,
Canal Companies, Shippers, Shipping Agents, General Carriers, etc. Ironfounders,
Brassfounders, Metal Merchants, Iron Manufacturers, Ironmongers, Engineers,
Machinists, Boiler Makers, Millwrights, Roofing, Bridge and Girder Makers, Colliery
Proprietors, etc. Timber Merchants, Builders, Contractors, Architects, Surveyors,
Auctioneers, Valuers, Brokers, Mill Owners and Manufacturers, Mill Furnishers,
Merchants, and General Wholesale Tradesmen. Also for the Apportionment of
Mileage Charges for Railway Traffic.*

"It is as easy of reference for any answer or any number of answers as a dictionary, and the
references are even more quickly made. For making up accounts or estimates the book must
prove invaluable to all who have any considerable quantity of calculations involving price and
measure in any combination to do."—*Engineer*.

Harben's Comprehensive Weight Calculator.

THE WEIGHT CALCULATOR. Being a Series of Tables
upon a New and Comprehensive Plan, exhibiting at One Reference the exact
Value of any Weight from 1 lb. to 15 tons, at 300 Progressive Rates, from 1d.
to 168s. per cwt., and containing 186,000 Direct Answers, which, with their
Combinations, consisting of a single addition (mostly to be performed at
sight), will afford an aggregate of 10,266,000 Answers; the whole being calculated and designed to ensure correctness and promote despatch. By HENRY
HARBEN, Accountant. Fourth Edition, carefully Corrected. Royal 8vo,
£1 5s. strongly half-bound.

"A practical and useful work of reference for men of business generally; it is the best of the
kind we have seen."—*Ironmonger*.
"Of priceless value to business men. It is a necessary book in all mercantile offices."—*Sheffield Independent*.

Harben's Comprehensive Discount Guide.

THE DISCOUNT GUIDE. Comprising several Series of
Tables for the use of Merchants, Manufacturers, Ironmongers, and others,
by which may be ascertained the exact Profit arising from any mode of using
Discounts, either in the Purchase or Sale of Goods, and the method of either
Altering a Rate of Discount or Advancing a Price, so as to produce, by one
operation, a sum that will realise any required profit after allowing one or
more Discounts: to which are added Tables of Profit or Advance from 1¼ to
90 per cent., Tables of Discount from 1½ to 98½ per cent., and Tables of Commission, &c., from ¼ to 10 per cent. By HENRY HARBEN, Accountant. New
Edition, Revised and Corrected. Demy 8vo, 544 pp., £1 5s. half-bound.

"A book such as this can only be appreciated by business men, to whom the saving of time
means saving of money. We have the high authority of Professor J. R. Young that the tables
throughout the work are constructed upon strictly accurate principles. The work is a model
of typographical clearness, and must prove of great value to merchants, manufacturers, and
general traders."—*British Trade Journal*.

New Wages Calculator.

TABLES OF WAGES at 54, 52, 50 and 48 Hours per Week,
Showing the Amounts of Wages from One-quarter-of-an-hour to Sixty-four
hours in each case at Rates of Wages advancing by One Shilling from 4s. to
55s. per week. By THOS. GARBUTT, Accountant. Square crown 8vo, 6s.
half-bound. [*Just published*.

Iron and Metal Trades' Calculator.

THE IRON AND METAL TRADES' COMPANION. For
expeditiously ascertaining the Value of any Goods bought or sold by Weight,
from 1s. per cwt. to 112s. per cwt., and from one farthing per pound to one
shilling per pound. By THOMAS DOWNIE. 396 pp., 9s. leather.
"A most useful set of tables; nothing like them before existed."—*Building News*.
"Although specially adapted to the iron and metal trades, the tables will be found useful in
every other business in which merchandise is bought and sold by weight."—*Railway News*.

"DIRECT CALCULATORS,"
By M. B. COTSWORTH, of Holgate, York.

QUICKEST AND MOST ACCURATE MEANS OF CALCULATION KNOWN.
ENSURE ACCURACY and SPEED WITH EASE, SAVE TIME and MONEY.
Accounts may be charged out or checked by these means in about **one**-third he time required by ordinary methods of calculation. These unrivalled "Calculators" have very clear and original contrivances for instantly finding the exact answer, by its fixed position, without even sighting the top or side of the page. They are varied in arrangement to suit the special need of each particular trade.

All the leading firms now use Calculators, even where they employ experts.

N.B.—Indicator letters in brackets should be quoted.

"*RAILWAY & TRADERS' CALCULATOR*" (**R. & T.**) 10s. 6d. Including Scale of Charges for Small Parcels by Merchandise Trains. "Direct Calculator"—the only Calculator published giving exact charge for Cwts., Qrs. and Lbs., together. "Calculating Tables" for every 1d. rate to 100s. per ton. "Wages Calculator." "Percentage Rates." "Grain, Flour, Ale, &c., Weight Calculators."

"*DIRECT CALCULATOR* (**I R**)" including all the above except "Calculating Tables." 7s.

"*DIRECT CALCULATOR* (**A**)" by ½d., 2s. each opening, exact pence to 40s. per ton. 5s.

"*DIRECT CALCULATOR* (**B**)" by 1d., 4s. each opening, exact pence to 40s. per ton. 4s. 6d.

"*DIRECT CALCULATOR* (**C**)" by 1d. (with Cwts. and Qrs. to nearest farthing), to 40s. per ton. 4s. 6d.

"*DIRECT CALCULATOR* (**Ds**)" by 1d. gradations. (Single Tons to 50 Tons, then by fifties to 1,000 Tons, with Cwts. values below in exact pence payable, fractions of ½d. and upwards being counted as 1d. 6s. 6d.

"*DIRECT CALCULATOR* (**D**)" has from 1,000 to 10,000 Tons in addition to the (**Ds**) Calculator. 7s. 6d.

"*DIRECT CALCULATOR* (**Es**)" by 1d. gradations. (As (**D**) to 1,000 Tons, with Cwts. and Qrs. values shown separately to the nearest farthing). 5s. 6d.

"*DIRECT CALCULATOR* (**E**)" has from 1,000 to 10,000 Tons in addition to the (**Es**) Calculator. 6s. 6d.

"*DIRECT CALCULATOR* (**F**)" by 1d., 2s. each opening, exact pence to 40s. per ton. 4s. 6d.

"*DIRECT CALCULATOR* (**G**)" by 1d., 1s. each opening; 6 in. by 9 in. Nearest ¼d. Indexed (**G I**) 3s. 6d. 2s. 6d.

"*DIRECT CALCULATOR* (**H**)" by 1d., 1s. each opening; 6 in. by 9 in. To exact pence. Indexed (**H I**) 3s. 6d. 2s. 6d.

"*DIRECT CALCULATOR* (**K**)" Showing Values of Tons, Cwts. and Qrs. in even pence (fractions of 1d. as 1d.), for the Retail Coal Trade. 4s. 6d.

"*RAILWAY AND TIMBER TRADES MEASURER AND CALCULATOR* (**T**)" (as prepared for the Railway Companies). The only book published giving true content of unequal sided and round timber by eighths of an inch, quarter girth, Weights from Cubic Feet—Standards, Superficial Feet, and Stone to Weights—Running Feet from lengths of Deals—Standard Multipliers—Timber Measures—Customs Regulations, &c. 3s. 6d.

AGRICULTURE, FARMING, GARDENING, etc.

Dr. Fream's New Edition of "The Standard Treatise on Agriculture."

THE COMPLETE GRAZIER, and FARMER'S and CATTLE-BREEDER'S ASSISTANT: A Compendium of Husbandry. Originally Written by WILLIAM YOUATT. Thirteenth Edition, entirely Re-written, considerably Enlarged, and brought up to the Present Requirements of Agricultural Practice, by WILLIAM FREAM, LL.D., Steven Lecturer in the University of Edinburgh, Author of "The Elements of Agriculture," &c. Royal 8vo, 1,100 pp., with over 450 Illustrations. £1 11s. 6d. strongly and handsomely bound.

EXTRACT FROM PUBLISHERS' ADVERTISEMENT.

"A treatise that made its original appearance in the first decade of the century, and that enters upon its Thirteenth Edition before the century has run its course, has undoubtedly established its position as a work of permanent value. . . The phenomenal progress of the last dozen years in the Practice and Science of Farming has rendered it necessary, however, that the volume should be re-written, . . . and for this undertaking the publishers were fortunate enough to secure the services of Dr. FREAM, whose high attainments in all matters pertaining to agriculture have been so emphatically recognised by the highest professional and official authorities. In carrying out his editorial duties, Dr. FREAM has been favoured with valuable contributions by Prof. J. WORTLEY AXE, Mr. E. BROWN, Dr. BERNARD DYER, Mr. W. J. MALDEN, Mr. R. H. REW, Prof. SHELDON, Mr. J. SINCLAIR, Mr. SANDERS SPENCER, and others.

"As regards the illustrations of the work, no pains have been spared to make them as representative and characteristic as possible, so as to be practically useful to the Farmer and Grazier."

SUMMARY OF CONTENTS.

BOOK I. ON THE VARIETIES, BREEDING, REARING, FATTENING, AND MANAGEMENT OF CATTLE.
BOOK II. ON THE ECONOMY AND MANAGEMENT OF THE DAIRY.
BOOK III. ON THE BREEDING, REARING, AND MANAGEMENT OF HORSES.
BOOK IV. ON THE BREEDING, REARING, AND FATTENING OF SHEEP.
BOOK V. ON THE BREEDING, REARING, AND FATTENING OF SWINE.
BOOK VI. ON THE DISEASES OF LIVE STOCK.

BOOK VII. ON THE BREEDING, REARING, AND MANAGEMENT OF POULTRY.
BOOK VIII. ON FARM OFFICES AND IMPLEMENTS OF HUSBANDRY.
BOOK IX. ON THE CULTURE AND MANAGEMENT OF GRASS LANDS.
BOOK X. ON THE CULTIVATION AND APPLICATION OF GRASSES, PULSE, AND ROOTS.
BOOK XI. ON MANURES AND THEIR APPLICATION TO GRASS LAND & CROPS.
BOOK XII. MONTHLY CALENDARS OF FARMWORK.

⁎ OPINIONS OF THE PRESS ON THE NEW EDITION.

"Dr. Fream is to be congratulated on the successful attempt he has made to give us a work which will at once become the standard classic of the farm practice of the country. We believe that it will be found that it has no compeer among the many works at present in existence. . . . The illustrations are admirable, while the frontispiece, which represents the well-known bull, New Year's Gift, bred by the Queen, is a work of art."—*The Times.*

"The book must be recognised as occupying the proud position of the most exhaustive work freference in the English language on the subject with which it deals."—*Athenæum.*

"The most comprehensive guide to modern farm practice that exists in the English language to-day. . . . The book is one that ought to be on every farm and in the library of every land-owner."—*Mark Lane Express.*

"In point of exhaustiveness and accuracy the work will certainly hold a pre-eminent and unique position among books dealing with scientific agricultural practice. It is, in fact, an agricultural library of itself."—*North British Agriculturist.*

"A compendium of authoritative and well-ordered knowledge on every conceivable branch of the work of the live stock farmer; probably without an equal in this or any other country."
Yorkshire Post.

British Farm Live Stock.

FARM LIVE STOCK OF GREAT BRITAIN. By ROBERT WALLACE, F.L.S., F.R.S.E., &c., Professor of Agriculture and Rural Economy in the University of Edinburgh. Third Edition, thoroughly Revised and considerably Enlarged. With over 120 Phototypes of Prize Stock. Demy 8vo, 384 pp., with 79 Plates and Maps, 12s. 6d. cloth.

"A really complete work on the history, breeds, and management of the farm stock of Great Britain, and one which is likely to find its way to the shelves of every country gentleman's library."—*The Times.*

"The latest edition of 'Farm Live Stock of Great Britain' is a production to be proud of, and its issue not the least of the services which its author has rendered to agricultural science."
Scottish Farmer,

"The book is very attractive . . . and we can scarcely imagine the existence of a farmer who would not like to have a copy of this beautiful work."—*Mark Lane Express.*

"A work which will long be regarded as a standard authority whenever a concise history and description of the breeds of live stock in the British Isles is required."—*Bell's Weekly Messenger.*

Dairy Farming.

BRITISH DAIRYING. A Handy Volume on the Work of the Dairy-Farm. For the Use of Technical Instruction Classes, Students in Agricultural Colleges, and the Working Dairy-Farmer. By Prof. J. P. SHELDON, ate Special Commissioner of the Canadian Government, Author of "Dairy Farming," &c. With numerous Illustrations. Crown 8vo, 2s. 6d. cloth.
"We confidently recommend it as a text-book on dairy farming."—*Agricultural Gazette.*
"Probably the best half-crown manual on dairy work that has yet been produced."—*North British Agriculturist.*
"It is the soundest little work we have yet seen on the subject."—*The Times.*

Dairy Manual.

MILK, CHEESE AND BUTTER: A Practical Handbook on their Properties and the Processes of their Production, including a Chapter on Cream and the Methods of its Separation from Milk. By JOHN OLIVER, late Principal of the Western Dairy Institute, Berkeley. With Coloured Plates and 200 Illusts. Crown 8vo, 7s. 6d. cloth.   [*Just published.*
"An exhaustive and masterly production. It may be cordially recommended to all students and practitioners of dairy science."—*N.B. Agriculturist.*
"We strongly recommend this very comprehensive and carefully-written book to dairy-farmers and students of dairying. It is a distinct acquisition to the library of the agriculturist."—*Agricultural Gazette.*

Agricultural Facts and Figures.

NOTE-BOOK OF AGRICULTURAL FACTS AND FIGURES FOR FARMERS AND FARM STUDENTS. By PRIMROSE McCONNELL, B.Sc. Fifth Edition. Royal 32mo, roan, gilt edges, with band, 4s.
"Literally teems with information, and we can cordially recommend it to all connected with agriculture."—*North British Agriculturist.*

Small Farming.

SYSTEMATIC SMALL FARMING; *or, The Lessons of my Farm.* Being an Introduction to Modern Farm Practice for Small Farmers. By R. SCOTT BURN. With numerous Illustrations, crown 8vo, 6s. cloth.
"This is the completest book of its class we have seen, and one which every amateur farmer will read with pleasure and accept as a guide."—*Field.*

Modern Farming.

OUTLINES OF MODERN FARMING. By R. SCOTT BURN. Soils, Manures, and Crops—Farming and Farming Economy—Cattle, Sheep, and Horses — Management of Dairy, Pigs, and Poultry — Utilisation of Town-Sewage, Irrigation, &c. Sixth Edition. In One Vol., 1,250 pp., half-bound, profusely Illustrated, 12s.
"The aim of the author has been to make his work at once comprehensive and trustworthy and he has succeeded to a degree which entitles him to much credit."—*Morning Advertiser.*

Agricultural Engineering.

FARM ENGINEERING, THE COMPLETE TEXT-BOOK OF. Comprising Draining and Embanking; Irrigation and Water Supply; Farm Roads, Fences, and Gates; Farm Buildings; Barn Implements and Machines; Field Implements and Machines; Agricultural Surveying, &c. By Prof. JOHN SCOTT. 1,150 pages, half-bound, with over 600 Illustrations, 12s.
"Written with great care, as well as with knowledge and ability. The author has done his work well; we have found him a very trustworthy guide wherever we have tested his statements. The volume will be of great value to agricultural students."—*Mark Lane Express.*

Agricultural Text-Book.

THE FIELDS OF GREAT BRITAIN: A Text-Book of Agriculture, adapted to the Syllabus of the Science and Art Department. For Elementary and Advanced Students. By HUGH CLEMENTS (Board of Trade). Second Edition, Revised, with Additions. 18mo, 2s. 6d. cloth.
"A most comprehensive volume, giving a mass of information."—*Agricultural Economist.*
"It is a long time since we have seen a book which has pleased us more, or which contains such a vast and useful fund of knowledge."—*Educational Times.*

Tables for Farmers, etc.

TABLES, MEMORANDA, AND CALCULATED RESULTS *for Farmers, Graziers, Agricultural Students, Surveyors, Land Agents, Auctioneers, etc.* With a New System of Farm Book-keeping. By SIDNEY FRANCIS. Third Edition, Revised. 272 pp., waistcoat-pocket size, 1s. 6d. leather.
"Weighing less than 1 oz., and occupying no more space than a match box, it contains a mass of facts and calculations which have never before, in such handy form, been obtainable. Every operation on the farm is dealt with. The work may be taken as thoroughly accurate, the whole of the tables having been revised by Dr. Fream. We cordially recommend it."—*Bell's Weekly Messenger.*

Artificial Manures and Foods.

FERTILISERS AND FEEDING STUFFS : Their Properties and Uses. A Handbook for the Practical Farmer. **By** BERNARD DYER, D.Sc. (Lond.) With the Text of the Fertilisers and Feeding Stuffs Act of 1893, the Regulations and Forms of the Board of Agriculture and Notes on the Act by A. J. DAVID, B.A., LL.M., of the Inner Temple, Barrister-at-Law. Crown 8vo, 120 pages, 1s. cloth. [*Just published.*]

"An excellent shillingsworth. Dr. Dyer has done farmers good service in placing at their disposa so much useful information in so intelligible a form."—*The Times.*

The Management of Bees.

BEES FOR PLEASURE AND PROFIT : A Guide to the Manipulation of Bees, the Production of Honey, and the General Management of the Apiary. By G. GORDON SAMSON. Crown 8vo, 1s. cloth.

"The intending bee-keeper will find exactly the kind of information required to enable him to make a successful start with his hives. The author is a thoroughly competent teacher, and his book may be commended."—*Morning Post.*

Farm and Estate Book-keeping.

BOOK-KEEPING FOR FARMERS & ESTATE OWNERS. A Practical Treatise, presenting, in Three Plans, a System adapted for all Classes of Farms. By JOHNSON M. WOODMAN, Chartered Accountant. Second Edition, Revised. Crown 8vo, 3s. 6d. cloth boards ; or 2s. 6d. cloth limp.

"The volume is a capital study of a most important subject."—*Agricultural Gazette.*

The young farmer, land agent, and surveyor will find Mr. Woodman's treatise **more than** repay its cost and study."— *Building News.*

Farm Account Book.

WOODMAN'S YEARLY FARM ACCOUNT BOOK. Giving a Weekly Labour Account and Diary, and showing the Income and Expenditure under each Department of Crops, Live Stock, Dairy, &c. &c. With Valuation, Profit and Loss Account, and Balance Sheet at the end of the Year. By JOHNSON M. WOODMAN, Chartered Accountant, Author of "Bookkeeping for Farmers." Folio, 7s. 6d. half bound. [*culture*]

"Contains every requisite form for keeping farm accounts readily and accurately."—*Agri-*

Early Fruits, Flowers, and Vegetables.

THE FORCING GARDEN ; or, How to Grow Early Fruits, Flowers, and Vegetables. With Plans and Estimates for Building Glasshouses, Pits, and Frames. By SAMUEL WOOD. Crown 8vo, 3s. 6d. cloth.

"A good book, and fairly fills a place that was in some degree vacant. The book is written with great care, and contains a great deal of valuable teaching."—*Gardeners' Magazine.*

Good Gardening.

A PLAIN GUIDE TO GOOD GARDENING ; or, How to Grow Vegetables, Fruits, and Flowers. By S. WOOD. Fourth Edition, with considerable Additions, &c., and numerous Illustrations. Crown 8vo, 3s. 6d. cl.

"A very good book, and one to be highly recommended as a practical guide. The practical directions are excellent."—*Athenaeum.*

"May be recommended to young gardeners, cottagers, and specially to amateurs, for the plain, simple, and trustworthy information it gives on common matters too often neglected."—*Gardeners' Chronicle.*

Gainful Gardening.

MULTUM-IN-PARVO GARDENING ; or, How to make One Acre of Land produce £620 a-year by the Cultivation of Fruits and Vegetables ; also, How to Grow Flowers in Three Glass Houses, so as to realise £176 per annum clear Profit. By SAMUEL WOOD, Author of "Good Gardening," &c. Fifth and Cheaper Edition, Revised, with Additions. Crown 8vo, 1s. sewed.

"We are bound to recommend it as not only suited to the case of the amateur and gentleman's gardener, but to the market grower."—*Gardeners' Magazine.*

Gardening for Ladies.

THE LADIES' MULTUM-IN-PARVO FLOWER GARDEN, and Amateurs' Complete Guide. With Illusts. By S. WOOD. Cr. 8vo, 3s. 6d. cl.

Receipts for Gardeners.

GARDEN RECEIPTS. Edited by CHARLES W. QUIN. 12mo, 1s. 6d. cloth limp.

Market Gardening.

MARKET AND KITCHEN GARDENING. By Contributors to "The Garden." Compiled by C. W. SHAW, late Editor of "Gardening Illustrated." 12mo 3s. 6d. cloth boards.

AUCTIONEERING, VALUING, LAND SURVEYING ESTATE AGENCY, etc.

Auctioneer's Assistant.

THE APPRAISER, AUCTIONEER, BROKER, HOUSE AND ESTATE AGENT AND VALUER'S POCKET ASSISTANT, for the Valuation for Purchase, Sale, or Renewal of Leases, Annuities and Reversions, and of property generally; with Prices for Inventories, &c. By JOHN WHEELER, Valuer, &c. Sixth Edition, Re-written and greatly extended by C. NORRIS, Surveyor, Valuer, &c. Royal 32mo, 5s. cloth.

" A neat and concise book of reference, containing an admirable and clearly-arranged list of prices for inventories, and a very practical guide to determine the value of furniture, &c."—*Standard.*

" Contains a large quantity of varied and useful information as to the valuation for purchase, sale, or renewal of leases, annuities and reversions, and of property generally, with prices for inventories, and a guide to determine the value of interior fittings and other effects."—*Builder.*

Auctioneering.

AUCTIONEERS: THEIR DUTIES AND LIABILITIES. A Manual of Instruction and Counsel for the Young Auctioneer. By ROBERT SQUIBBS, Auctioneer. Second Edition, Revised and partly Re-written. Demy 8vo, 12s. 6d. cloth.

*** OPINIONS OF THE PRESS.

" The standard text-book on the topics of which it treats."—*Athenæum.*

" The work is one of general excellent character, and gives much information in a compendious and satisfactory form."—*Builder.*

" May be recommended as giving a great deal of information on the law relating to auctioneers, in a very readable form."—*Law Journal.*

" Auctioneers may be congratulated on having so pleasing a writer to minister to their special needs."—*Solicitors' Journal.*

" Every auctioneer ought to possess a copy of this excellent work."—*Ironmonger.*

" Of great value to the profession. . . . We readily welcome this book from the fact that it treats the subject in a manner somewhat new to the profession."—*Estates Gazette.*

Inwood's Estate Tables.

TABLES FOR THE PURCHASING OF ESTATES, Freehold, Copyhold, or Leasehold; Annuities, Advowsons, etc., and for the Renewing of Leases held under Cathedral Churches, Colleges, or other Corporate bodies for Terms of Years certain, and for Lives; also for Valuing Reversionary Estates, Deferred Annuities, Next Presentations, &c.; together with SMART'S Five Tables of Compound Interest, and an Extension of the same to Lower and Intermediate Rates. By W. INWOOD. 24th Edition, with considerable Additions, and new and valuable Tables of Logarithms for the more Difficult Computations of the Interest of Money, Discount, Annuities, &c., by M. FEDOR THOMAN, of the Société Crédit Mobilier of Paris. Crown 8vo, 8s. cloth.

" Those interested in the purchase and sale of estates, and in the adjustment of compensation cases, as well as in transactions in annuities, life insurances, &c., will find the present edition of eminent service."—*Engineering.*

" 'Inwood's Tables' still maintain a most enviable reputation. The new issue has been enriched by large additional contributions by M. Fedor Thoman, whose carefully arranged Tables cannot fail to be of the utmost utility."—*Mining Journal.*

Agricultural Valuer's Assistant.

THE AGRICULTURAL VALUER'S ASSISTANT. A Practical Handbook on the Valuation of Landed Estates; including Rules and Data for Measuring and Estimating the Contents, Weights, and Values of Agricultural Produce and Timber, and the Values of Feeding Stuffs, Manures, and Labour; with Forms of Tenant-Right-Valuations, Lists of Local Agricultural Customs, Scales of Compensation under the Agricultural Holdings Act, &c. &c. By TOM BRIGHT, Agricultural Surveyor. Second Edition, much Enlarged. Crown 8vo, 5s. cloth.

" Full of tables and examples in connection with the valuation of tenant-right, estates, labour, contents, and weights of timber, and farm produce of all kinds."—*Agricultural Gazette.*

" An eminently practical handbook, full of practical tables and data of undoubted interest and value to surveyors and auctioneers in preparing valuations of all kinds."—*Farmer.*

Plantations and Underwoods.

POLE PLANTATIONS AND UNDERWOODS: A Practical Handbook on Estimating the Cost of Forming, Renovating, Improving, and Grubbing Plantations and Underwoods, their Valuation for Purposes of Transfer, Rental, Sale, or Assessment. By TOM BRIGHT, Author of "The Agricultural Valuer's Assistant," &c. Crown 8vo, 3s. 6d. cloth.

" To valuers, foresters and agents it will be a welcome aid."—*North British Agriculturist.*

" Well calculated to assist the valuer in the discharge of his duties, and of undoubted interest and use both to surveyors and auctioneers in preparing valuations of all kinds."—*Kent Herald.*

Hudson's Land Valuer's Pocket-Book.

THE LAND VALUER'S BEST ASSISTANT: Being Tables on a very much Improved Plan, for Calculating the Value of Estates. With Tables for reducing Scotch, Irish, and Provincial Customary Acres to Statute Measure, &c. By R. HUDSON, C.E. New Edition. Royal 32mo, 4s. leather.
"Of incalculable value to the country gentleman and professional man."—*Farmers' Journal.*

Ewart's Land Improver's Pocket-Book.

THE LAND IMPROVER'S POCKET-BOOK OF FORMULÆ, TABLES, and MEMORANDA required in any Computation relating to the Permanent Improvement of Landed Property. By JOHN EWART, Surveyor. Second Edition. Royal 32mo, 4s. leather.

Complete Agricultural Surveyor's Pocket-Book.

THE LAND VALUER'S AND LAND IMPROVER'S COM-PLETE POCKET-BOOK. Being of the above Two Works bound together. Leather, with strap, 7s. 6d.

House Property.

HANDBOOK OF HOUSE PROPERTY. A Popular and Practical Guide to the Purchase, Mortgage, Tenancy, and Compulsory Sale of Houses and Land, including the Law of Dilapidations and Fixtures; with Examples of all kinds of Valuations, Useful Information on Building, and Suggestive Elucidations of Fine Art. By E. L. TARBUCK, Architect and Surveyor. Fifth Edition, Enlarged. 12mo, 5s. cloth.
"The advice is thoroughly practical."—*Law Journal.*
"For all who have dealings with house property, this is an indispensable guide."—*Decoration.*
"Carefully brought up to date, and much improved by the addition of a division on fine art. . . . A well-written and thoughtful work."—*Land Agent's Record.*

LAW AND MISCELLANEOUS.

Pocket-Book for Sanitary Officials.

THE HEALTH OFFICER'S POCKET-BOOK: A Guide to Sanitary Practice and Law. For Medical Officers of Health, Sanitary Inspectors, Members of Sanitary Authorities, &c. By EDWARD F. WILLOUGHBY, M.D. (Lond.), &c., Author of "Hygiene and Public Health." Fcap. 8vo, 7s. 6d. cloth, red edges, rounded corners. [*Just published.*
"A mine of condensed information of a pertinent and useful kind on the various subjects of which it treats. The matter seems to have been carefully compiled and arranged for facility of reference, and it is well illustrated by diagrams and woodcuts. The different subjects are succinctly but fully and scientifically dealt with."—*The Lancet.*
"An excellent publication, dealing with the scientific, technical and legal matters connected with the duties of medical officers of health and sanitary inspectors. The work is replete with information."—*Local Government Journal.*

Journalism.

MODERN JOURNALISM. A Handbook of Instruction and Counsel for the Young Journalist. By JOHN B. MACKIE, Fellow of the Institute of Journalists. Crown 8vo, 2s. cloth. [*Just published.*
"This invaluable guide to journalism is a work which all aspirants to a journalistic career will read with advantage."—*Journalist.*

Private Bill Legislation and Provisional Orders.

HANDBOOK FOR THE USE OF SOLICITORS AND EN-GINEERS Engaged in Promoting Private Acts of Parliament and Provisional Orders, for the Authorization of Railways, Tramways, Gas and Water Works, &c. By L. LIVINGSTON MACASSEY, of the Middle Temple, Barrister-at-Law, M.Inst.C.E. 8vo, 25s. cloth.

Law of Patents.

PATENTS FOR INVENTIONS, AND HOW TO PROCURE THEM. Compiled for the Use of Inventors, Patentees and others. By G. G. M. HARDINGHAM, Assoc.Mem.Inst.C.E., &c. Demy 8vo, 1s. 6d. cloth.

Labour Disputes.

CONCILIATION AND ARBITRATION IN LABOUR DIS-PUTES: A Historical Sketch and Brief Statement of the Present Position of the Question at Home and Abroad. By J. S. JEANS, Author of "England's Supremacy," &c. Crown 8vo, 200 pp., 2s. 6d. cloth. [*Just published.*
"Mr. Jeans is well qualified to write on this subject, both by his previous books and by his practical experience as an arbitrator."—*The Times.*

A Complete Epitome of the Laws of this Country.

EVERY MAN'S OWN LAWYER: A Handy-Book of the Principles of Law and Equity. With A Concise Dictionary of Legal Terms. By A Barrister. Thirty-second Edition, carefully Revised, and including New Acts of Parliament of 1894. Comprising the *Local Government Act,* 1894 (establishing District and Parish Councils); *Finance Act,* 1894 (imposing the New Death Duties); *Merchant Shipping Act,* 1894; *Prevention of Cruelty to Children Act,* 1894; *Building Societies Act,* 1894; *Notice of Accidents Act,* 1894; *Sale of Goods Act,* 1893; *Voluntary Conveyances Act,* 1893; *Married Women's Property Act,* 1893; *Trustee Act,* 1893; *Fertiliser and Feeding Stuffs Act,* 1893; *Betting and Loans (Infants) Act,* 1892; *Shop Hours Act,* 1892; *Small Holdings Act,* 1892; and many other important new Acts. Crown 8vo, 750 pp., price 6s. 8d. (saved at every consultation!), strongly bound in cloth. [*Just published.*

** *The Book will be found to comprise (amongst other matter)—*

The Rights and Wrongs of Individuals—Landlord and Tenant—Vendors and Purchasers—Leases and Mortgages—Principal and Agent—Partnership and Companies—Masters, Servants, and Workmen—Contracts and Agreements—Borrowers, Lenders, and Sureties—Sale and Purchase of Goods—Cheques, Bills, and Notes—Bills of Sale—Bankruptcy—Railway and Shipping Law—Life, Fire, and Marine Insurance—Accident and Fidelity Insurance—Criminal Law—Parliamentary Elections—County Councils—District Councils—Parish Councils—Municipal Corporations—Libel and Slander—Public Health and Nuisances—Copyright, Patents, Trade Marks—Husband and Wife—Divorce—Infancy—Custody of Children—Trustees and Executors—Clergy, Churchwardens, etc.—Game Laws and Sporting—Innkeepers—Horses and Dogs—Taxes and Death Duties—Forms of Agreements, Wills, Codicils, Notices, etc.

☞ *The object of this work is to enable those who consult it to help themselves to the law; and thereby to dispense, as far as possible, with professional assistance and advice. There are many wrongs and grievances which persons submit to from time to time through not knowing how or where to apply for redress; and many persons have as great a dread of a lawyer's office as of a lion's den. With this book at hand it is believed that many a Six-and-Eightpence may be saved; many a wrong redressed; many a right reclaimed; many a law suit avoided; and many an evil abated. The work has established itself as the standard legal adviser of all classes, and has also made a reputation for itself as a useful book of reference for lawyers residing at a distance from law libraries, who are glad to have at hand a work embodying recent decisions and enactments.*

** Opinions of the Press.

"A complete code of English Law, written in plain language, which all can understand. ould be in the hands of every business man, and all who wish to abolish lawyers' bills. . . ." *Weekly Times.*

" A useful and concise epitome of the law, compiled with considerable care."—*Law Magazine.*

"A complete digest of the most useful facts which constitute English law."—*Globe.*

"Admirably done, admirably arranged, and admirably cheap."—*Leeds Mercury.*

" A concise, cheap and complete epitome of the English law. So plainly written that he who runs may read, and he who reads may understand."—*Figaro.*

"The latest edition of this popular book ought to be in every business establishment, and on every library table."—*Sheffield Post.*

" A complete epitome of the law; thoroughly intelligible to non-professional readers." *Bell's Life.*

Legal Guide for Pawnbrokers.

THE LAW OF LOANS AND PLEDGES. With Statutes and a Digest of Cases. By H. C. Folkard, Esq., Barrister-at-Law. Fcap, 8vo, 3s. 6d. cloth.

The Law of Contracts.

LABOUR CONTRACTS: A Popular Handbook on the Law of Contracts for Works and Services. By David Gibbons. Fourth Edition, Appendix of Statutes by T. F. Uttley, Solicitor. Fcap. 8vo, 3s. 6d. cloth.

The Factory Acts.

SUMMARY OF THE FACTORY AND WORKSHOP ACTS (1878-1891). For the Use of Manufacturers and Managers. By Emile Garcke and J. M. Fells. (Reprinted from "Factory Accounts.") Crown 8vo, 6d. sewed.

OGDEN, SMALE AND CO. LIMITED, PRINTERS, GREAT SAFFRON HILL. E.C.

𝔚eale's 𝔕udimentary 𝔖eries.

LONDON, 1862.
THE PRIZE MEDAL
Was awarded to the Publishers of

WEALE'S SERIES
RUDIMENTARY SCIENTIFIC, EDUCATIONAL, AND CLASSICAL.

Comprising nearly Three Hundred and Fifty distinct works in almost every department of Science, Art, and Education, recommended to the notice of Engineers, Architects, Builders, Artisans, and Students generally, as well as to those interested in Workmen's Libraries, Literary and Scientific Institutions, Colleges, Schools, Science Classes, &c., &c.

☞ "WEALE'S SERIES includes Text-Books on almost every branch of Science and Industry, comprising such subjects as Agriculture, Architecture and Building, Civil Engineering, Fine Arts, Mechanics and Mechanical Engineering, Physical and Chemical Science, and many miscellaneous Treatises. The whole are constantly undergoing revision, and new editions, brought up to the latest discoveries in scientific research, are constantly issued. The prices at which they are sold are as low as their excellence is assured."—*American Literary Gazette.*

"Amongst the literature of technical education, WEALE'S SERIES has ever enjoyed a high reputation, and the additions being made by Messrs. CROSBY LOCKWOOD & SON render the series more complete, and bring the information upon the several subjects down to the present time."—*Mining Journal.*

"It is not too much to say that no books have ever proved more popular with, or more useful to, young engineers and others than the excellent treatises comprised in WEALE'S SERIES."—*Engineer.*

"The excellence of WEALE'S SERIES is now so well appreciated, that it would be wasting our space to enlarge upon their general usefulness and value."—*Builder.*

"The volumes of WEALE'S SERIES form one of the best collections of elementary technical books in any language."—*Architect.*

"WEALE'S SERIES has become a standard as well as an unrivalled collection of treatises in all branches of art and science."—*Public Opinion.*

PHILADELPHIA, 1876.
THE PRIZE MEDAL
Was awarded to the Publishers for
Books : Rudimentary, Scientific,
"WEALE'S SERIES," ETC.

CROSBY LOCKWOOD & SON,
7, STATIONERS' HALL COURT, LUDGATE HILL, LONDON, E.C.

⁎⁎⁎ Catalogues post free on application.

WEALE'S RUDIMENTARY SCIENTIFIC SERIES.

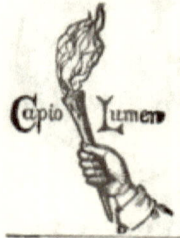

. The volumes of this Series are freely Illustrated with Woodcuts, or otherwise, where requisite. Throughout the following List it must be understood that the books are bound in limp cloth, unless otherwise stated; *but the volumes marked with a ‡ may also be had strongly bound in cloth boards for 6d. extra.*

N.B.—In ordering from this List it is recommended, as a means of facilitating business and obviating error, to quote the numbers affixed to the volumes, as well as the titles and prices.

CIVIL ENGINEERING, SURVEYING, ETC.

No.

31. *WELLS AND WELL-SINKING.* By John Geo. Swindell, A.R.I.B.A., and G. R. Burnell, C.E. Revised Edition. With a New Appendix on the Qualities of Water. Illustrated. 2s.

35. *THE BLASTING AND QUARRYING OF STONE*, for Building and other Purposes. By Gen. Sir J. Burgoyne, Bart. 1s. 6d.

43. *TUBULAR, AND OTHER IRON GIRDER BRIDGES*, particularly describing the Britannia and Conway Tubular Bridges. By G. Drysdale Dempsey, C.E. Fourth Edition. 2s.

44. *FOUNDATIONS AND CONCRETE WORKS*, with Practical Remarks on Footings, Sand, Concrete, Béton, Pile-driving, Caissons, and Cofferdams, &c. By E. Dobson. Seventh Edition. 1s. 6d.

60. *LAND AND ENGINEERING SURVEYING.* By T. Baker, C.E. Fifteenth Edition, revised by Professor J. R. Young. 2s.‡

80*. *EMBANKING LANDS FROM THE SEA.* With examples and Particulars of actual Embankments, &c. By J. Wiggins, F.G.S. 2s.

81. *WATER WORKS*, for the Supply of Cities and Towns. With a Description of the Principal Geological Formations of England as influencing Supplies of Water, &c. By S. Hughes, C.E. New Edition. 4s.‡

118. *CIVIL ENGINEERING IN NORTH AMERICA*, a Sketch of. By David Stevenson, F.R.S.E., &c. Plates and Diagrams. 3s.

167. *IRON BRIDGES, GIRDERS, ROOFS, AND OTHER WORKS.* By Francis Campin, C.E. 2s. 6d.‡

197. *ROADS AND STREETS.* By H. Law, C.E., revised and enlarged by D. K. Clark, C.E., including pavements of Stone, Wood, Asphalte, &c. 4s. 6d.‡

203. *SANITARY WORK IN THE SMALLER TOWNS AND IN VILLAGES.* By C. Slagg, A.M.I.C.E. Revised Edition. 3s.‡

212. *GAS-WORKS, THEIR CONSTRUCTION AND ARRANGEMENT*; and the Manufacture and Distribution of Coal Gas. Originally written by Samuel Hughes, C.E. Re-written and enlarged by William Richards, C.E. Eighth Edition, with important additions. 5s. 6d.‡

213. *PIONEER ENGINEERING.* A Treatise on the Engineering Operations connected with the Settlement of Waste Lands in New Countries. By Edward Dobson, Assoc. Inst. C.E. 4s. 6d.‡

216. *MATERIALS AND CONSTRUCTION;* A Theoretical and Practical Treatise on the Strains, Designing, and Erection of Works of Construction. By Francis Campin, C.E. Second Edition, revised. 3s.‡

419. *CIVIL ENGINEERING.* By Henry Law, M.Inst. C.E. Including Hydraulic Engineering by Geo. R. Burnell, M.Inst. C.E. Seventh Edition, revised, with large additions by D Kinnear Clark, M.Inst. C.E. 6s. 6d., Cloth boards, 7s. 6d.

268. *THE DRAINAGE OF LANDS, TOWNS, & BUILDINGS.* By G. D. Dempsey, C.E. Revised, with large Additions on Recent Practice in Drainage Engineering, by D. Kinnear Clark, M.I.C.E. Second Edition, Corrected. 4s. 6d.‡

☞ *The ‡ indicates that these vols. may be had strongly bound at 6d. extra.*

LONDON : CROSBY LOCKWOOD AND SON.

MECHANICAL ENGINEERING, ETC.

The ‡ indicates that these vols. may be had strongly bound at 6d. extra.

MINING, METALLURGY, ETC.

4. *MINERALOGY*, Rudiments of; a concise View of the General Properties of Minerals. By A. RAMSAY, F.G.S., F.R.G.S., &c. Fourth Edition, revised and enlarged. Illustrated. 3s. 6d.‡

117. *SUBTERRANEOUS SURVEYING*, with and without the Magnetic Needle. By T. FENWICK and T. BAKER, C.E. Illustrated. 2s. 6d. ‡

135. *ELECTRO-METALLURGY;* Practically Treated. By ALEXANDER WATT. Tenth Edition, enlarged, with additional Illustrations, and including the most recent Processes. 3s. 6d.‡

172. *MINING TOOLS*, Manual of. For the Use of Mine Managers, Agents, Students, &c. By WILLIAM MORGANS. 2s. 6d.

172*. *MINING TOOLS, ATLAS* of Engravings to Illustrate the above, containing 235 Illustrations, drawn to Scale. 4to. 4s. 6d.

176. *METALLURGY OF IRON*. Containing History of Iron Manufacture, Methods of Assay, and Analyses of Iron Ores, Processes of Manufacture of Iron and Steel, &c., By H. BAUERMAN, F.G.S. Sixth Edition, revised and enlarged. 5s.‡

180. *COAL AND COAL MINING*. By the late Sir WARINGTON W. SMYTH, M.A., F.R.S. Seventh Edition, revised. 3s. 6d.‡

195. *THE MINERAL SURVEYOR AND VALUER'S COMPLETE GUIDE*. By W. LINTERN, M.E. Third Edition, including Magnetic and Angular Surveying. With Four Plates. 3s. 6d.‡

214. *SLATE AND SLATE QUARRYING*, Scientific, Practical, and Commercial. By D. C. DAVIES, F.G.S., Mining Engineer, &c. 3s.‡

264. *A FIRST BOOK OF MINING AND QUARRYING*, with the Sciences connected therewith, for Primary Schools and Self Instruction. By J H. COLLINS, F.G.S. Second Edition, with additions. 1s. 6d.

ARCHITECTURE, BUILDING, ETC.

16. *ARCHITECTURE—ORDERS*—The Orders and their Æsthetic Principles. By W. H. LEEDS. Illustrated. 1s. 6d.

17. *ARCHITECTURE—STYLES*—The History and Description of the Styles of Architecture of Various Countries, from the Earliest to the Present Period. By T. TALBOT BURY, F.R.I.B.A., &c. Illustrated. 2s.
*** ORDERS AND STYLES OF ARCHITECTURE, *in One Vol.*, 3s. 6d.

18. *ARCHITECTURE—DESIGN*—The Principles of Design in Architecture, as deducible from Nature and exemplified in the Works of the Greek and Gothic Architects. By E. L. GARBETT, Architect. Illustrated. 2s.6d.
. *The three preceding Works, in One handsome Vol., half bound, entitled* "MODERN ARCHITECTURE," *price* 6s.*.

22. *THE ART OF BUILDING*, Rudiments of. General Principles of Construction, Materials used in Building, Strength and Use of Materials, Working Drawings, Specifications, and Estimates. By E. DOBSON, 2s.‡

25. *MASONRY AND STONECUTTING:* Rudimentary Treatise on the Principles of Masonic Projection and their application to Construction. By EDWARD DOBSON, M.R.I.B.A., &c. 2s. 6d.‡

42. *COTTAGE BUILDING*. By C. BRUCE ALLEN, Architect. Eleventh Edition, revised and enlarged. With a Chapter on Economic Cottages for Allotments, by EDWARD E. ALLEN, C.E. 2s.

45. *LIMES, CEMENTS, MORTARS, CONCRETES, MASTICS,* ‡PLASTERING, &c. By G. R. BURNELL, C.E Fourteenth Edition. 1s. 6d

57. *WARMING AND VENTILATION*. An Exposition of the General Principles as applied to Domestic and Public Buildings, Mines, Lighthouses, Ships, &c. By C. TOMLINSON, F.R.S., &c. Illustrated. 3s.

111. *ARCHES, PIERS, BUTTRESSES, &c.:* Experimental Essays on the Principles of Construction. By W. BLAND. Illustrated. 1s. 6d.

The ‡ *indicates that these vols. may be had strongly bound at 6d. extra.*

LONDON : CROSBY LOCKWOOD AND SON,

Architecture, Building, etc., *continued.*

116. *THE ACOUSTICS OF PUBLIC BUILDINGS;* or, The Principles of the Science of Sound applied to the purposes of the Architect and Builder. By T. ROGER SMITH, M.R.I.B.A., Architect. Illustrated. 1s. 6d.

127. *ARCHITECTURAL MODELLING IN PAPER,* the Art of. By T. A. RICHARDSON, Architect. Illustrated. 1s. 6d.

128. *VITRUVIUS — THE ARCHITECTURE OF MARCUS VITRUVIUS POLLO.* In Ten Books. Translated from the Latin by JOSEPH GWILT, F.S.A., F.R.A.S. With 23 Plates. 5s.

130. *GRECIAN ARCHITECTURE,* An Inquiry into the Principles of Beauty in; with an Historical View of the Rise and Progress of the Art in Greece. By the EARL OF ABERDEEN. 1s.

⁎⁎⁎ *The two preceding Works in One handsome Vol., half bound, entitled* "ANCIENT ARCHITECTURE," *price 6s.*

132. *THE ERECTION OF DWELLING-HOUSES.* Illustrated by a Perspective View, Plans, Elevations, and Sections of a pair of Semi-detached Villas, with the Specification, Quantities, and Estimates, &c. By S. H. BROOKS. New Edition, with Plates. 2s. 6d.‡

156. *QUANTITIES & MEASUREMENTS* in Bricklayers', Masons', Plasterers', Plumbers', Painters', Paperhangers', Gilders', Smiths', Carpenters' and Joiners' Work. By A. C. BEATON, Surveyor. Ninth Edition. 1s. 6d.

175. *LOCKWOOD'S BUILDER'S PRICE BOOK FOR* 1895. A Comprehensive Handbook of the Latest Prices and Data for Builders, Architects, Engineers, and Contractors. Re-constructed, Re-written, and further Enlarged. By FRANCIS T. W. MILLER, A.R.I.B.A. 800 pages. 4s., cloth boards.

182. *CARPENTRY AND JOINERY* — THE ELEMENTARY PRINCIPLES OF CARPENTRY. Chiefly composed from the Standard Work of THOMAS TREDGOLD, C.E. With a TREATISE ON JOINERY by E. WYNDHAM TARN, M.A. Fifth Edition, Revised. 3s. 6d.‡

182*. *CARPENTRY AND JOINERY.* ATLAS of 35 Plates to accompany the above. With Descriptive Letterpress. 4to. 6s.

185. *THE COMPLETE MEASURER;* the Measurement of Boards, Glass, &c.; Unequal-sided, Square-sided, Octagonal-sided, Round Timber and Stone, and Standing Timber, &c. By RICHARD HORTON. Fifth Edition. 4s.; strongly bound in leather, 5s.

187. *HINTS TO YOUNG ARCHITECTS.* By G. WIGHTWICK. New Edition. By G. H. GUILLAUME. Illustrated. 3s. 6d.‡

188. *HOUSE PAINTING, GRAINING, MARBLING, AND SIGN WRITING:* with a Course of Elementary Drawing for House-Painters, Sign-Writers, &c., and a Collection of Useful Receipts. By ELLIS A. DAVIDSON. Sixth Edition. With Coloured Plates. 5s. cloth limp; 6s. cloth boards.

189. *THE RUDIMENTS OF PRACTICAL BRICKLAYING.* In Six Sections: General Principles; Arch Drawing, Cutting, and Setting; Pointing; Paving, Tiling, Materials; Slating and Plastering; Practical Geometry, Mensuration, &c. By ADAM HAMMOND. Eighth Edition. 1s. 6d.

191. *PLUMBING.* A Text-Book to the Practice of the Art or Craft of the Plumber. With Chapters upon House Drainage and Ventilation. Sixth Edition. With 380 Illustrations. By W. P. BUCHAN. 3s. 6d.‡

192. *THE TIMBER IMPORTER'S, TIMBER MERCHANT'S,* and BUILDER'S STANDARD GUIDE. By R. E. GRANDY. 2s.

206. *A BOOK ON BUILDING,* Civil and Ecclesiastical, including CHURCH RESTORATION. With the Theory of Domes and the Great Pyramid, &c. By Sir EDMUND BECKETT, Bart., LL.D., Q.C., F.R.A.S. 4s. 6d.‡

226. *THE JOINTS MADE AND USED BY BUILDERS* in the Construction of various kinds of Engineering and Architectural Works. By WYVILL J. CHRISTY, Architect. With upwards of 160 Engravings on Wood. 3s.‡

228. *THE CONSTRUCTION OF ROOFS OF WOOD AND IRON.* By E. WYNDHAM TARN, M.A., Architect. Third Edition, revised. 1s. 6d.

☞ *The ‡ indicates that these vols. may be had strongly bound at 6d. extra.*

Architecture, Building, etc., *continued.*

229. *ELEMENTARY DECORATION:* as applied to the Interior and Exterior Decoration of Dwelling-Houses, &c. By J. W. FACEY. 2s.

257. *PRACTICAL HOUSE DECORATION.* A Guide to the Art of Ornamental Painting. By JAMES W. FACEY. 2s. 6d.

₊ *The two preceding Works, in One handsome Vol., half-bound, entitled* "HOUSE DECORATION, ELEMENTARY AND PRACTICAL," *price* 5s.

230. *A PRACTICAL TREATISE ON HANDRAILING.* Showing New and Simple Methods. By G. COLLINGS. Second Edition, Revised, including A TREATISE ON STAIRBUILDING. Plates. 2s. 6d.

247. *BUILDING ESTATES:* a Rudimentary Treatise on the Development, Sale, Purchase, and General Management of Building Land. By FOWLER MAITLAND, Surveyor. Second Edition, revised. 2s.

248. *PORTLAND CEMENT FOR USERS.* By HENRY FAIJA, Assoc. M. Inst. C.E. Third Edition, corrected. Illustrated. 2s.

252. *BRICKWORK:* a Practical Treatise, embodying the General and Higher Principles of Bricklaying, Cutting and Setting, &c. By F. WALKER. Third Edition, Revised and Enlarged. 1s. 6d.

23. *THE PRACTICAL BRICK AND TILE BOOK.* Comprising :
189. BRICK AND TILE MAKING, by E. DOBSON, A.I.C.E.; PRACTICAL BRICKLAY-
265. ING, by A HAMMOND; BRICKCUTTING AND SETTING, by A. HAMMOND. 534 pp. with 270 Illustrations. 6s. Strongly half-bound.

253. *THE TIMBER MERCHANT'S, SAW-MILLER'S, AND IMPORTER'S FREIGHT-BOOK AND ASSISTANT.* By WM. RICHARDSON. With Additions by M. POWIS BALE, A.M.Inst.C.E. 3s.‡

258. *CIRCULAR WORK IN CARPENTRY AND JOINERY.* A Practical Treatise on Circular Work of Single and Double Curvature. By GEORGE COLLINGS. Second Edition, 2s. 6d.

259. *GAS FITTING:* A Practical Handbook treating of every Description of Gas Laying and Fitting. By J. BLACK. Second Edition, 2s. 6d.‡

261. *SHORING AND ITS APPLICATION:* A Handbook for the Use of Students. By GEORGE H. BLAGROVE. 1s. 6d.

265. *THE ART OF PRACTICAL BRICK CUTTING & SETTING.* By ADAM HAMMOND. With 90 Engravings. 1s. 6d.

267. *THE SCIENCE OF BUILDING:* An Elementary Treatise on the Principles of Construction. By E. WYNDHAM TARN, M.A. Lond. Third Edition, Revised and Enlarged. 3s. 6d.‡

271. *VENTILATION:* a Text-book to the Practice of the Art of Ventilating Buildings By W. P. BUCHAN, R.P., Sanitary Engineer, Author of "Plumbing," &c. 3s. 6d.‡

272. *ROOF CARPENTRY;* Practical Lessons in the Framing of Wood Roofs. For the Use of Working Carpenters. By GEO. COLLINGS, Author of "Handrailing and Stairbuilding," &c. 2s.

273. *THE PRACTICAL PLASTERER:* A Compendium of Plain and Ornamental Plaster Work. By WILFRED KEMP. 2s.

SHIPBUILDING, NAVIGATION, ETC.

51. *NAVAL ARCHITECTURE.* An Exposition of the Elementary Principles. By J. PEAKE. Fifth Edition, with Plates. 3s. 6d.‡

53*. *SHIPS FOR OCEAN & RIVER SERVICE,* Elementary and Practical Principles of the Construction of. By H. A. SOMMERFELDT. 1s. 6d.

53**. *AN ATLAS OF ENGRAVINGS* to Illustrate the above. Twelve large folding plates. Royal 4to, cloth. 7s. 6d.

54. *MASTING, MAST-MAKING, AND RIGGING OF SHIPS,* Also Tables of Spars, Rigging, Blocks; Chain, Wire, and Hemp Ropes &c., relative to every class of vessels. By ROBERT KIPPING, N.A. 2s.

☞ *The ‡ indicates that these vols. may be had strongly bound at 6d. extra.*

Shipbuilding, Navigation, Marine Engineering, etc., *cont.*

54*. *IRON SHIP-BUILDING.* With Practical Examples and Details. By JOHN GRANTHAM, C.E. Fifth Edition. 4s.

55. *THE SAILOR'S SEA BOOK:* a Rudimentary Treatise on Navigation. By JAMES GREENWOOD, B.A. With numerous Woodcuts and Coloured Plates. New and enlarged edition. By W. H. ROSSER. 2s. 6d.‡

80. *MARINE ENGINES AND STEAM VESSELS.* By ROBERT MURRAY, C.E. Eighth Edition, thoroughly Revised, with Additions by the Author and by GEORGE CARLISLE, C.E. 4s. 6d. limp; 5s. cloth boards.

83*bis.* *THE FORMS OF SHIPS AND BOATS.* By W. BLAND. Ninth Edition, with numerous Illustrations and Models. 1s. 6d.

99. *NAVIGATION AND NAUTICAL ASTRONOMY*, in Theory and Practice. By Prof. J. R. YOUNG. New Edition. 2s. 6d.

106. *SHIPS' ANCHORS*, a Treatise on. By G. COTSELL, N.A. 1s. 6d.

149. *SAILS AND SAIL-MAKING.* With Draughting, and the Centre of Effort of the Sails; Weights and Sizes of Ropes; Masting, Rigging, and Sails of Steam Vessels, &c. 13th Edition. By R. KIPPING, N.A., 2s. 6d.‡

155. *ENGINEER'S GUIDE TO THE ROYAL & MERCANTILE* NAVIES. By a PRACTICAL ENGINEER. Revised by D. F. M'CARTHY. 3s.

55 & 204. *PRACTICAL NAVIGATION.* Consisting of The Sailor's Sea-Book. By JAMES GREENWOOD and W. H. ROSSER. Together with the requisite Mathematical and Nautical Tables for the Working of the Problems. By H. LAW, C.E., and Prof. J. R. YOUNG. 7s. Half-bound.

AGRICULTURE, GARDENING, ETC.

61*. *A COMPLETE READY RECKONER FOR THE ADMEA-*SUREMENT OF LAND, &c. By A. ARMAN. Fourth Edition, revised and extended by C. NORRIS, Surveyor, Valuer, &c. 2s.

131. *MILLER'S, CORN MERCHANT'S, AND FARMER'S* READY RECKONER. Second Edition, with a Price List of Modern Flour-Mill Machinery, by W. S. HUTTON, C.E. 2s.

140. *SOILS, MANURES, AND CROPS.* (Vol. 1. OUTLINES OF MODERN FARMING.) By R. SCOTT BURN. Woodcuts. 2s.

141. *FARMING & FARMING ECONOMY*, Notes, Historical and Practical, on. (Vol. 2. OUTLINES OF MODERN FARMING.) By R. SCOTT BURN. 3s.

142. *STOCK; CATTLE, SHEEP, AND HORSES.* (Vol. 3. OUTLINES OF MODERN FARMING.) By R. SCOTT BURN. Woodcuts. 2s. 6d.

145. *DAIRY, PIGS, AND POULTRY*, Management of the. By R. SCOTT BURN. (Vol. 4. OUTLINES OF MODERN FARMING.) 2s.

146. *UTILIZATION OF SEWAGE, IRRIGATION, AND* RECLAMATION OF WASTE LAND. (Vol. 5. OUTLINES OF MODERN FARMING.) By R. SCOTT BURN. Woodcuts. 2s. 6d.

₊ *Nos. 140-1-2-5-6, in One Vol., handsomely half-bound, entitled "*OUTLINES OF MODERN FARMING.*" By* ROBERT SCOTT BURN. *Price 12s.*

177. *FRUIT TREES*, The Scientific and Profitable Culture of. From the French of DU PREUIL. Revised by GEO. GLENNY. 187 Woodcuts. 3s. 6d.

198. *SHEEP; THE HISTORY, STRUCTURE, ECONOMY, AND* DISEASES OF. By W. C. SPOONER, M.R.V.C., &c. Fifth Edition, enlarged, including Specimens of New and Improved Breeds. 3s. 6d.‡

201. *KITCHEN GARDENING MADE EASY.* By GEORGE M. F. GLENNY. Illustrated. 1s. 6d.‡

207. *OUTLINES OF FARM MANAGEMENT, and the Organi-*zation of Farm Labour. By R. SCOTT BURN. 2s. 6d.‡

208. *OUTLINES OF LANDED ESTATES MANAGEMENT.* By R. SCOTT BURN. 2s. 6d.

₊ *Nos. 207 & 208 in One Vol., handsomely half-bound, entitled "*OUTLINES OF LANDED ESTATES AND FARM MANAGEMENT.*" By* R. SCOTT BURN. *Price 6s.*

☞ *The ‡ indicates that these vols. may be had strongly bound at 6d. extra.*

Agriculture, Gardening, etc., *continued.*

209. *THE TREE PLANTER AND PLANT PROPAGATOR.*
A Practical Manual on the Propagation of Forest Trees, Fruit Trees, Flowering Shrubs, Flowering Plants, &c. By SAMUEL WOOD. 2s.

210. *THE TREE PRUNER.* A Practical Manual on the Pruning of Fruit Trees, including also their Training and Renovation; also the Pruning of Shrubs, Climbers, and Flowering Plants. By SAMUEL WOOD. 1s. 6d.

*** *Nos.* 209 *&* 210 *in One Vol., handsomely half-bound, entitled* "THE TREE PLANTER, PROPAGATOR, AND PRUNER." By SAMUEL WOOD. *Price* 3s. 6d.

218. *THE HAY AND STRAW MEASURER :* Being New Tables for the Use of Auctioneers, Valuers, Farmers, Hay and Straw Dealers, &c. By JOHN STEELE. Fifth Edition. 2s.

222. *SUBURBAN FARMING.* The Laying-out and Cultivation of Farms, adapted to the Produce of Milk, Butter, and Cheese, Eggs, Poultry and Pigs. By Prof. JOHN DONALDSON and R. SCOTT BURN. 3s. 6d.‡

231. *THE ART OF GRAFTING AND BUDDING.* By CHARLES BALTET. With Illustrations. 2s. 6d.‡

232. *COTTAGE GARDENING ;* or, Flowers, Fruits, and Vegetables for Small Gardens. By E. HOBDAY. 1s. 6d.

233. *GARDEN RECEIPTS.* Edited by CHARLES W. QUIN. 1s. 6d.

234. *MARKET AND KITCHEN GARDENING.* By C. W. SHAW, late Editor of "Gardening Illustrated." 3s.‡

239. *DRAINING AND EMBANKING.* A Practical Treatise, embodying the most recent experience in the Application of Improved Methods. By JOHN SCOTT, late Professor of Agriculture and Rural Economy at the Royal Agricultural College, Cirencester. With 68 Illustrations. 1s. 6d.

240. *IRRIGATION AND WATER SUPPLY.* A Treatise on Water Meadows, Sewage Irrigation, and Warping; the Construction of Wells, Ponds, and Reservoirs, &c. By Prof. JOHN SCOTT. With 34 Illus. 1s. 6d.

241. *FARM ROADS, FENCES, AND GATES.* A Practical Treatise on the Roads, Tramways, and Waterways of the Farm; the Principles of Enclosures; and the different kinds of Fences, Gates, and Stiles. By Professor JOHN SCOTT. With 75 Illustrations. 1s. 6d.

242. *FARM BUILDINGS.* A Practical Treatise on the Buildings necessary for various kinds of Farms, their Arrangement and Construction, with Plans and Estimates. By Prof. JOHN SCOTT. With 105 Illus. 2s.

243. *BARN IMPLEMENTS AND MACHINES.* A Practical Treatise on the Application of Power to the Operations of Agriculture; and on various Machines used in the Threshing-barn, in the Stock-yard, and in the Dairy, &c. By Prof. J. SCOTT. With 123 Illustrations. 2s.

244. *FIELD IMPLEMENTS AND MACHINES.* A Practical Treatise on the Varieties now in use, with Principles and Details of Construction, their Points of Excellence, and Management. By Professor JOHN SCOTT. With 138 Illustrations. 2s.

245. *AGRICULTURAL SURVEYING.* A Practical Treatise on Land Surveying, Levelling, and Setting-out; and on Measuring and Estimating Quantities, Weights, and Values of Materials, Produce, Stock, &c. By Prof. JOHN SCOTT. With 62 Illustrations. 1s. 6d.

*** *Nos.* 239 *to* 245 *in One Vol., handsomely half-bound, entitled* "THE COMPLETE TEXT-BOOK OF FARM ENGINEERING." By Professor JOHN SCOTT. *Price* 12s.

250. *MEAT PRODUCTION.* A Manual for Producers, Distributors, &c. By JOHN EWART. 2s. 6d.‡

266. *BOOK-KEEPING FOR FARMERS & ESTATE OWNERS.* By J. M. WOODMAN, Chartered Accountant. 2s. 6d. cloth limp; 3s. 6d. cloth boards.

☞ *The* ‡ *indicates that these vols may be had strongly bound at* 6d. *extra.*

LONDON : CROSBY LOCKWOOD AND SON.

MATHEMATICS, ARITHMETIC, ETC.

32. *MATHEMATICAL INSTRUMENTS*, a Treatise on; Their Construction, Adjustment, Testing, and Use concisely Explained. By J. F. HEATHER, M.A. Fourteenth Edition, revised, with additions, by A. T. WALMISLEY, M.I.C.E., Fellow of the Surveyors' Institution. Original Edition, in 1 vol., Illustrated. 2s.‡

*** *In ordering the above, be careful to say, " Original Edition" (No. 32), to distinguish it from the Enlarged Edition in 3 vols. (Nos. 168-9-70.)*

76. *DESCRIPTIVE GEOMETRY*, an Elementary Treatise on; with a Theory of Shadows and of Perspective, extracted from the French of G. MONGE. To which is added, a description of the Principles and Practice of Isometrical Projection. By J. F. HEATHER, M.A. With 14 Plates. 2s.

178. *PRACTICAL PLANE GEOMETRY:* giving the Simplest Modes of Constructing Figures contained in one Plane and Geometrical Construction of the Ground. By J. F. HEATHER, M.A. With 215 Woodcuts. 2s.

83. *COMMERCIAL BOOK-KEEPING.* With Commercial Phrases and Forms in English, French, Italian, and German. By JAMES HADDON, M.A., Arithmetical Master of King's College School, London. 1s. 6d.

84. *ARITHMETIC*, a Rudimentary Treatise on: with full Explanations of its Theoretical Principles, and numerous Examples for Practice. By Professor J. R. YOUNG. Twelfth Edition. 1s. 6d.

84*. A KEY to the above, containing Solutions in full to the Exercises, together with Comments, Explanations, and Improved Processes, for the Use of Teachers and Unassisted Learners. By J. R. YOUNG. 1s. 6d.

85. *EQUATIONAL ARITHMETIC*, applied to Questions of Interest, Annuities, Life Assurance, and General Commerce; with various Tables by which all Calculations may be greatly facilitated. By W. HIPSLEY. 2s.

86. *ALGEBRA*, the Elements of. By JAMES HADDON, M.A. With Appendix, containing miscellaneous Investigations, and a Collection of Problems in various parts of Algebra. 2s.

86*. A KEY AND COMPANION to the above Book, forming an extensive repository of Solved Examples and Problems in Illustration of the various Expedients necessary in Algebraical Operations. By J. R. YOUNG. 1s. 6d.

88. *EUCLID*, THE ELEMENTS OF: with many additional Propositions
89. and Explanatory Notes: to which is prefixed, an Introductory Essay on Logic. By HENRY LAW, C.E. 2s. 6d.‡

*** *Sold also separately, viz. :—*

88. EUCLID, The First Three Books. By HENRY LAW, C.E. 1s. 6d.
89. EUCLID, Books 4, 5, 6, 11, 12. By HENRY LAW, C.E. 1s. 6d.

90. *ANALYTICAL GEOMETRY AND CONIC SECTIONS* By JAMES HANN. A New Edition, by Professor J. R. YOUNG. 2s.‡

91. *PLANE TRIGONOMETRY*, the Elements of. By JAMES HANN, formerly Mathematical Master of King's College, London. 1s. 6d.

92. *SPHERICAL TRIGONOMETRY*, the Elements of. By JAMES HANN. Revised by CHARLES H. DOWLING, C.E. 1s.

*** *Or with "The Elements of Plane Trigonometry," in One Volume, 2s. 6d.*

93. *MENSURATION AND MEASURING.* With the Mensuration and Levelling of Land for the Purposes of Modern Engineering. By T. BAKER, C.E. New Edition by E. NUGENT, C.E. Illustrated. 1s. 6d.

101. *DIFFERENTIAL CALCULUS*, Elements of the. By W. S. B. WOOLHOUSE, F.R.A.S., &c. 1s. 6d.

102. *INTEGRAL CALCULUS*, Rudimentary Treatise on the. By HOMERSHAM COX, B.A. Illustrated. 1s.

136. *ARITHMETIC*, Rudimentary, for the Use of Schools and Self-Instruction. By JAMES HADDON, M.A. Revised by A. ARMAN. 1s. 6d.
137. A KEY TO HADDON'S RUDIMENTARY ARITHMETIC. By A. ARMAN. 1s. 6d.

☞ *The ‡ indicates that these vols. may be had strongly bound at 6d. extra.*

7, STATIONERS' HALL COURT, LUDGATE HILL, E.C.

Mathematics, Arithmetic, etc., *continued.*

168. *DRAWING AND MEASURING INSTRUMENTS.* Including—I. Instruments employed in Geometrical and Mechanical Drawing, and in the Construction, Copying, and Measurement of Maps and Plans. II. Instruments used for the purposes of Accurate Measurement, and for Arithmetical Computations. By J. F. HEATHER, M.A. Illustrated. 1s. 6d

169. *OPTICAL INSTRUMENTS.* Including (more especially) Telescopes, Microscopes, and Apparatus for producing copies of Maps and Plans by Photography. By J. F. HEATHER, M.A. Illustrated. 1s. 6d.

170. *SURVEYING AND ASTRONOMICAL INSTRUMENTS.* Including—I. Instruments Used for Determining the Geometrical Features of a portion of Ground. II. Instruments Employed in Astronomical Observations. By J. F. HEATHER, M.A. Illustrated. 1s. 6d.

₊ *The above three volumes form an enlargement of the Author's original work,* "*Mathematical Instruments.*" *(See No. 32 in the Series.)*

168.
169. } *MATHEMATICAL INSTRUMENTS.* By J. F. HEATHER, M.A. Enlarged Edition, for the most part entirely re-written. The 3 Parts as
170. above, in One thick Volume. With numerous Illustrations. 4s. 6d.‡

158. *THE SLIDE RULE, AND HOW TO USE IT;* containing full, easy, and simple Instructions to perform all Business Calculations with unexampled rapidity and accuracy. By CHARLES HOARE, C.E. Sixth Edition. With a Slide Rule in tuck of cover. 2s. 6d.‡

196. *THEORY OF COMPOUND INTEREST AND ANNUITIES;* with Tables of Logarithms for the more Difficult Computations of Interest, Discount, Annuities, &c. By FÉDOR THOMAN. Fourth Edition. 4s.‡

199. *THE COMPENDIOUS CALCULATOR;* or, Easy and Concise Methods of Performing the various Arithmetical Operations required in Commercial and Business Transactions; together with Useful Tables. By D. O GORMAN. Twenty-seventh Edition, carefully revised by C. NORRIS. 2s. 6d., cloth limp; 3s. 6d., strongly half-bound in leather.

204. *MATHEMATICAL TABLES*, for Trigonometrical, Astronomical, and Nautical Calculations; to which is prefixed a Treatise on Logarithms. By HENRY LAW, C.E. Together with a Series of Tables for Navigation and Nautical Astronomy. By Prof. J. R. YOUNG. New Edition. 4s.

204*. *LOGARITHMS.* With Mathematical Tables for Trigonometrical, Astronomical, and Nautical Calculations. By HENRY LAW, M.Inst.C.E. New and Revised Edition. (Forming part of the above Work). 3s.

221. *MEASURES, WEIGHTS, AND MONEYS OF ALL NATIONS,* and an Analysis of the Christian, Hebrew, and Mahometan Calendars. By W. S. B. WOOLHOUSE, F.R.A.S., F.S.S. Seventh Edition. 2s. 6d.‡

227. *MATHEMATICS AS APPLIED TO THE CONSTRUCTIVE ARTS.* Illustrating the various processes of Mathematical Investigation, by means of Arithmetical and Simple Algebraical Equations and Practical Examples. By FRANCIS CAMPIN, C.E. Third Edition. 3s.‡

PHYSICAL SCIENCE, NATURAL PHILOSOPHY, ETC.

1. *CHEMISTRY.* By Professor GEORGE FOWNES, F.R.S. With an Appendix on the Application of Chemistry to Agriculture. 1s.

2. *NATURAL PHILOSOPHY,* Introduction to the Study of. By C. TOMLINSON. Woodcuts. 1s. 6d.

6. *MECHANICS,* Rudimentary Treatise on. By CHARLES TOMLINSON. Illustrated. 1s. 6d.

7. *ELECTRICITY;* showing the General Principles of Electrical Science, and the purposes to which it has been applied. By Sir W. SNOW HARRIS, F.R.S., &c. With Additions by R. SABINE, C.E., F.S.A. 1s. 6d.

7*. *GALVANISM.* By Sir W. SNOW HARRIS. New Edition by ROBERT SABINE, C.E., F.S.A. 1s. 6d.

8. *MAGNETISM;* being a concise Exposition of the General Principles of Magnetical Science. By Sir W. SNOW HARRIS. New Edition, revised by H. M. NOAD, Ph.D. With 165 Woodcuts. 3s. 6d.‡

The ‡ indicates that these vols. may be had strongly bound at 6d. extra.

Physical Science, Natural Philosophy, etc., *continued.*

11. *THE ELECTRIC TELEGRAPH;* its History and Progress; with Descriptions of some of the Apparatus. By R. SABINE, C.E., F.S.A. 3s.

12. *PNEUMATICS,* including Acoustics and the Phenomena of Wind Currents, for the Use of Beginners By CHARLES TOMLINSON, F.R.S. Fourth Edition, enlarged. Illustrated. 1s. 6d.

72. *MANUAL OF THE MOLLUSCA;* a Treatise on Recent and Fossil Shells. By Dr. S. P. WOODWARD, A.L.S. Fourth Edition. With Plates and 300 Woodcuts. 7s. 6d., cloth.

96. *ASTRONOMY.* By the late Rev. ROBERT MAIN, M.A. Third Edition, by WILLIAM THYNNE LYNN, B.A., F.R.A.S. 2s.

97. *STATICS AND DYNAMICS,* the Principles and Practice of; embracing also a clear development of Hydrostatics, Hydrodynamics, and Central Forces. By T. BAKER, C.E. Fourth Edition. 1s. 6d.

173. *PHYSICAL GEOLOGY,* partly based on Major-General PORT-LOCK'S "Rudiments of Geology." By RALPH TATE, A.L.S., &c. Woodcuts. 2s.

174. *HISTORICAL GEOLOGY,* partly based on Major-General PORTLOCK'S "Rudiments." By RALPH TATE, A.L.S., &c. Woodcuts. 2s. 6d.

173 & 174. *RUDIMENTARY TREATISE ON GEOLOGY,* Physical and Historical. Partly based on Major-General PORTLOCK'S "Rudiments of Geology." By RALPH TATE, A.L.S., F.G.S., &c. In One Volume. 4s. 6d.‡

183 & 184. *ANIMAL PHYSICS,* Handbook of. By Dr. LARDNER, D.C.L., formerly Professor of Natural Philosophy and Astronomy in University College, Lond. With 520 Illustrations. In One Vol. 7s. 6d., cloth boards. ***** *Sold also in Two Parts, as follows :—*

183. ANIMAL PHYSICS. By Dr. LARDNER. Part I., Chapters I.—VII. 4s.

184. ANIMAL PHYSICS. By Dr. LARDNER. Part II., Chapters VIII.—XVIII. 3s.

269. *LIGHT:* an Introduction to the Science of Optics, for the Use of Students of Architecture, Engineering, and other Applied Sciences. By E. WYNDHAM TARN, M.A. 1s. 6d.

FINE ARTS.

20. *PERSPECTIVE FOR BEGINNERS.* Adapted to Young Students and Amateurs in Architecture, Painting, &c. By GEORGE PYNE. 2s.

40 *GLASS STAINING, AND THE ART OF PAINTING ON GLASS.* From the German of Dr. GESSERT and EMANUEL OTTO FROMBERG. With an Appendix on THE ART OF ENAMELLING. 2s 6d.

69. *MUSIC,* A Rudimentary and Practical Treatise on. With numerous Examples. By CHARLES CHILD SPENCER. 2s. 6d.

71. *PIANOFORTE,* The Art of Playing the. With numerous Exercises & Lessons from the Best Masters. By CHARLES CHILD SPENCER. 1s. 6d.

69-71. *MUSIC & THE PIANOFORTE.* In one vol. Half bound, 5s.

181. *PAINTING POPULARLY EXPLAINED,* including Fresco, Oil, Mosaic, Water Colour, Water-Glass, Tempera, Encaustic, Miniature, Painting on Ivory, Vellum, Pottery, Enamel, Glass, &c. With Historical Sketches of the Progress of the Art by THOMAS JOHN GULLICK, assisted by JOHN TIMBS, F.S.A. Sixth Edition, revised and enlarged. 5s.‡

186. *A GRAMMAR OF COLOURING,* applied to Decorative Painting and the Arts. By GEORGE FIELD. New Edition, enlarged and adapted to the Use of the Ornamental Painter and Designer. By ELLIS A. DAVIDSON. With two new Coloured Diagrams, &c. 3s.‡

246. *A DICTIONARY OF PAINTERS, AND HANDBOOK FOR PICTURE AMATEURS;* including Methods of Painting, Cleaning, Relining and Restoring, Schools of Painting, &c. With Notes on the Copyists and Imitators of each Master. By PHILIPPE DARYL. 2s. 6d.‡

The ‡ indicates that these vols. may be had strongly bound at 6d. extra.

INDUSTRIAL AND USEFUL ARTS.

23. *BRICKS AND TILES*, Rudimentary Treatise on the Manufacture of. By E. DOBSON, M.R.I.B.A. Illustrated, 3s.‡

67. *CLOCKS, WATCHES, AND BELLS*, a Rudimentary Treatise on. By Sir EDMUND BECKETT, LL.D., Q.C. Seventh Edition, revised and enlarged. 4s. 6d. limp; 5s. 6d. cloth boards.

83**. *CONSTRUCTION OF DOOR LOCKS*. Compiled from the Papers of A. C. HOBBS, and Edited by CHARLES TOMLINSON. F.R.S. 2s. 6d.

162. *THE BRASS FOUNDER'S MANUAL;* Instructions for Modelling, Pattern-Making, Moulding, Turning, Filing, Burnishing, Bronzing, &c. With copious Receipts, &c. By WALTER GRAHAM. 2s.‡

205. *THE ART OF LETTER PAINTING MADE EASY.* By J. G. BADENOCH. Illustrated with 12 full-page Engravings of Examples. 1s. 6d.

215. *THE GOLDSMITH'S HANDBOOK*, containing full Instructions for the Alloying and Working of Gold. By GEORGE E. GEE. 3s.‡

225. *THE SILVERSMITH'S HANDBOOK*, containing full Instructions for the Alloying and Working of Silver. By GEORGE E. GEE. 3s.‡

. *The two preceding Works, in One handsome Vol., half-bound, entitled "*THE GOLDSMITH'S & SILVERSMITH'S COMPLETE HANDBOOK,*" 7s.*

249. *THE HALL-MARKING OF JEWELLERY PRACTICALLY CONSIDERED.* By GEORGE E. GEE. 3s.‡

224. *COACH BUILDING*, A Practical Treatise, Historical and Descriptive. By J. W. BURGESS. 2s. 6d.‡

235. *PRACTICAL ORGAN BUILDING.* By W. E. DICKSON, M.A., Precentor of Ely Cathedral. Illustrated. 2s. 6d.‡

262. *THE ART OF BOOT AND SHOEMAKING.* By JOHN BEDFORD LENO. Numerous Illustrations. Third Edition. 2s.

263. *MECHANICAL DENTISTRY:* A Practical Treatise on the Construction of the Various Kinds of Artificial Dentures, with Formulæ, Tables, Receipts, &c. By CHARLES HUNTER. Third Edition. 3s.‡

270. *WOOD ENGRAVING:* A Practical and Easy Introduction to the Study of the Art. By W. N. BROWN. 1s. 6d.

MISCELLANEOUS VOLUMES.

36. *A DICTIONARY OF TERMS used in ARCHITECTURE, BUILDING, ENGINEERING, MINING, METALLURGY, ARCHÆOLOGY, the FINE ARTS, &c.* By JOHN WEALE. Sixth Edition. Revised by ROBERT HUNT, F.R.S. Illustrated. 5s. limp; 6s. cloth boards.

50. *LABOUR CONTRACTS.* A Popular Handbook on the Law of Contracts for Works and Services. By DAVID GIBBONS. Fourth Edition, Revised, with Appendix of Statutes by T. F. UTTLEY, Solicitor, 3s. 6d. cloth.

112. *MANUAL OF DOMESTIC MEDICINE.* By R. GOODING, B.A., M.D. A Family Guide in all Cases of Accident and Emergency 2s.

112*. *MANAGEMENT OF HEALTH.* A Manual of Home and Personal Hygiene. By the Rev. JAMES BAIRD, B.A. 1s.

150. *LOGIC*, Pure and Applied. By S. H. EMMENS. 1s. 6d.

153. *SELECTIONS FROM LOCKE'S ESSAYS ON THE HUMAN UNDERSTANDING.* With Notes by S. H. EMMENS. 1s. 6d.

154. *GENERAL HINTS TO EMIGRANTS.* 2s.

157. *THE EMIGRANT'S GUIDE TO NATAL.* By R. MANN. 2s.

193. *HANDBOOK OF FIELD FORTIFICATION.* By Major W. W. KNOLLYS, F.R.G.S. With 163 Woodcuts. 3s.‡

194. *THE HOUSE MANAGER:* Being a Guide to Housekeeping, Practical Cookery, Pickling and Preserving, Household Work, Dairy Management, &c. By AN OLD HOUSEKEEPER. 3s. 6d.‡

194, *HOUSE BOOK (The)*. Comprising :—I. THE HOUSE MANAGER.
112 & By an OLD HOUSEKEEPER. II. DOMESTIC MEDICINE. By R. GOODING, M.D.
112*. III. MANAGEMENT OF HEALTH. By J. BAIRD. In One Vol., half-bound, 6s.

☞ *The ‡ indicates that these vols. may be had strongly bound at 6d. extra.*

LONDON : CROSBY LOCKWOOD AND SON.

EDUCATIONAL AND CLASSICAL SERIES.

HISTORY.

1. **England, Outlines of the History of;** more especially with reference to the Origin and Progress of the English Constitution. By WILLIAM DOUGLAS HAMILTON, F.S.A., of Her Majesty's Public Record Office. 4th Edition, revised. 5s.; cloth boards, 6s.

5. **Greece, Outlines of the History of;** in connection with the Rise of the Arts and Civilization in Europe. By W. DOUGLAS HAMILTON, of University College, London, and EDWARD LEVIEN, M.A., of Balliol College, Oxford. 2s. 6d.; cloth boards, 3s. 6d.

7. **Rome, Outlines of the History of:** from the Earliest Period to the Christian Era and the Commencement of the Decline of the Empire. By EDWARD LEVIEN, of Balliol College, Oxford. Map, 2s. 6d.; cl. bds. 3s. 6d.

9. **Chronology of History, Art, Literature, and Progress,** from the Creation of the World to the Present Time. The Continuation by W. D. HAMILTON, F.S.A. 3s.; cloth boards, 3s. 6d.

ENGLISH LANGUAGE AND MISCELLANEOUS.

11. **Grammar of the English Tongue, Spoken and Written.** With an Introduction to the Study of Comparative Philology. By HYDE CLARKE, D.C.L. Fifth Edition. 1s. 6d.

12. **Dictionary of the English Language, as Spoken and Written.** Containing above 100,000 Words. By HYDE CLARKE, D.C.L. 3s. 6d.; cloth boards, 4s. 6d.; complete with the GRAMMAR, cloth bds., 5s. 6d.

48. **Composition and Punctuation,** familiarly Explained for those who have neglected the Study of Grammar. By JUSTIN BRENAN. 18th Edition. 1s. 6d.

49. **Derivative Spelling-Book:** Giving the Origin of Every Word from the Greek, Latin, Saxon, German, Teutonic, Dutch, French, Spanish, and other Languages; with their present Acceptation and Pronunciation. By J. ROWBOTHAM, F.R.A.S. Improved Edition. 1s. 6d.

51. **The Art of Extempore Speaking:** Hints for the Pulpit, the Senate, and the Bar. By M. BAUTAIN, Vicar-General and Professor at the Sorbonne. Translated from the French. 8th Edition, carefully corrected. 2s. 6d.

53. **Places and Facts in Political and Physical Geography,** for Candidates in Examinations. By the Rev. EDGAR RAND, B.A. 1s.

54. **Analytical Chemistry,** Qualitative and Quantitative, a Course of. To which is prefixed, a Brief Treatise upon Modern Chemical Nomenclature and Notation. By WM. W. PINK and GEORGE E. WEBSTER. 2s.

THE SCHOOL MANAGERS' SERIES OF READING BOOKS,

Edited by the Rev. A. R. GRANT, Rector of Hitcham, and Honorary Canon of Ely; formerly H.M. Inspector of Schools.

INTRODUCTORY PRIMER, 3d.

	s.	d.					s.	d.
FIRST STANDARD	0	6	FOURTH STANDARD	.	.	.	1	2
SECOND „	0	10	FIFTH „	.	.	.	1	6
THIRD „	1	0	SIXTH „	.	.	.	1	6

LESSONS FROM THE BIBLE. Part I. Old Testament. 1s.

LESSONS FROM THE BIBLE. Part II. New Testament, to which is added THE GEOGRAPHY OF THE BIBLE, for very young Children. By Rev. C. THORNTON FORSTER. 1s. 2d. *₊* Or the Two Parts in One Volume. 2s.

7, STATIONERS' HALL COURT, LUDGATE HILL, E.C.

FRENCH.

24. **French Grammar.** With Complete and Concise Rules on the Genders of French Nouns. By G. L. STRAUSS, Ph.D. 1s. 6d.
25. **French–English Dictionary.** Comprising a large number of New Terms used in Engineering, Mining, &c. By ALFRED ELWES. 1s. 6d.
26. **English–French Dictionary.** By ALFRED ELWES. 2s.
25,26. **French Dictionary** (as above). Complete, in One Vol., 3s.; cloth boards, 3s. 6d. *₊* Or with the GRAMMAR, cloth boards, 4s. 6d.
47. **French and English Phrase Book**: containing Introductory Lessons, with Translations, several Vocabularies of Words a Collection of suitable Phrases, and Easy Familiar Dialogues. 1s. 6d.

GERMAN.

39. **German Grammar.** Adapted for English Students, from Heyse's Theoretical and Practical Grammar, by Dr. G. L. STRAUSS. 1s. 6d.
40. **German Reader:** A Series of Extracts, carefully culled from the most approved Authors of Germany; with Notes, Philological and Explanatory. By G. L. STRAUSS, Ph.D. 1s.
41-43. **German Triglot Dictionary.** By N. E. S. A. HAMILTON. In Three Parts. Part I. German-French-English. Part II. English-German-French. Part III. French-German-English. 3s., or cloth boards, 4s.
41-43 **German Triglot Dictionary** (as above), together with German
& 39. Grammar (No. 39), in One Volume, cloth boards, 5s.

ITALIAN.

27. **Italian Grammar,** arranged in Twenty Lessons, with a Course of Exercises. By ALFRED ELWES. 1s. 6d.
28. **Italian Triglot Dictionary,** wherein the Genders of all the Italian and French Nouns are carefully noted down. By ALFRED ELWES. Vol. 1. Italian-English-French. 2s. 6d.
30. **Italian Triglot Dictionary.** By A. ELWES. Vol. 2. English-French-Italian. 2s. 6d.
32. **Italian Triglot Dictionary.** By ALFRED ELWES. Vol. 3. French-Italian-English. 2s. 6d.
28,30, **Italian Triglot Dictionary** (as above). In One Vol., 7s. 6d.
32. Clo h boards.

SPANISH AND PORTUGUESE.

34. **Spanish Grammar,** in a Simple and Practical Form. With a Course of Exercises. By ALFRED ELWES. 1s. 6d.
35. **Spanish–English and English–Spanish Dictionary.** Including a large number of Technical Terms used in Mining, Engineering, &c. with the proper Accents and the Gender of every Noun. By ALFRED ELWES 4s.; cloth boards, 5s. *₊* Or with the GRAMMAR, cloth boards, 6s.
55. **Portuguese Grammar,** in a Simple and Practical Form. With a Course of Exercises. By ALFRED ELWES. 1s. 6d.
56. **Portuguese–English and English–Portuguese Dictionary.** Including a large number of Technical Terms used in Mining, Engineering, &c., with the proper Accents and the Gender of every Noun. By ALFRED ELWES. Third Edition, Revised, 5s.; cloth boards, 6s. *₊* Or with the GRAMMAR, cloth boards, 7s.

HEBREW.

46*. **Hebrew Grammar.** By Dr. BRESSLAU. 1s. 6d.
44. **Hebrew and English Dictionary,** Biblical and Rabbinical; containing the Hebrew and Chaldee Roots of the Old Testament Post-Rabbinical Writings. By Dr. BRESSLAU. 6s.
46. **English and Hebrew Dictionary.** By Dr. BRESSLAU. 3s.
44,46. **Hebrew Dictionary** (as above), in Two Vols., complete, with
46*. the GRAMMAR, cloth boards, 12s.

LONDON : CROSBY LOCKWOOD AND SON,

LATIN.

19. **Latin Grammar.** Containing the Inflections and Elementary Principles of Translation and Construction. By the Rev. THOMAS GOODWIN, M.A., Head Master of the Greenwich Proprietary School. 1s. 6d.

20. **Latin-English Dictionary.** By the Rev. THOMAS GOODWIN, M.A. 2s.

22. **English-Latin Dictionary;** together with an Appendix of French and Italian Words which have their origin from the Latin. By the Rev. THOMAS GOODWIN, M.A. 1s. 6d.

20,22. **Latin Dictionary** (as above). Complete in One Vol., 3s. 6d. cloth boards, 4s. 6d. *.* Or with the GRAMMAR, cloth boards, 5s. 6d.

LATIN CLASSICS. With Explanatory Notes in English.

1. **Latin Delectus.** Containing Extracts from Classical Authors, with Genealogical Vocabularies and Explanatory Notes, by H. YOUNG. 1s. 6d.

2. **Cæsaris** Commentarii de Bello Gallico. Notes, and a Geographical Register for the Use of Schools, by H. YOUNG. 2s.

3. **Cornelius Nepos.** With Notes. By H. YOUNG. 1s.

4. **Virgilii** Maronis Bucolica et Georgica. With Notes on the Bucolics by W. RUSHTON, M.A., and on the Georgics by H. YOUNG. 1s. 6d.

5. **Virgilii** Maronis Æneis. With Notes, Critical and Explanatory, by H. YOUNG. **New** Edition, revised and improved With copious Additional Notes by Rev. T. H. L. LEARY, D.C.L., formerly Scholar of Brasenose College, Oxford. 3s.

5*. ——— Part 1. Books i.—vi., 1s. 6d.
5**. ——— Part 2. Books vii.—xii., 2s.

6. **Horace;** Odes, Epode, and Carmen Sæculare. Notes by H. YOUNG. 1s. 6d.

7. **Horace;** Satires, Epistles, and Ars Poetica. Notes by W. BROWNRIGG SMITH, M.A., F.R.G.S. 1s. 6d.

8. **Sallustii** Crispi Catalina et Bellum Jugurthinum. Notes, Critical and Explanatory, by W. M. DONNE, B.A., Trin. Coll., Cam. 1s. 6d.

9. **Terentii** Andria et Heautontimorumenos. With Notes, Critical and Explanatory, by the Rev. JAMES DAVIES, M.A. 1s. 6d.

10. **Terentii** Adelphi, Hecyra, **Phormio.** Edited, with Notes, **Critical** and Explanatory, by the Rev. JAMES DAVIES, M.A. 2s.

11. **Terentii** Eunuchus, Comœdia. Notes, by Rev. J. DAVIES, M.A. 1s. 6d.

12. **Ciceronis** Oratio pro Sexto Roscio Amerino. Edited, with an Introduction, Analysis, and Notes, Explanatory and Critical, by the Rev. JAMES DAVIES, M.A. 1s. 6d.

13. **Ciceronis** Orationes in Catilinam, Verrem, et pro Archia. With Introduction, Analysis, and Notes, Explanatory and Critical, by Rev. T. H. L. LEARY, D.C.L. formerly Scholar of Brasenose College, Oxford. 1s. 6d.

14. **Ciceronis** Cato Major, Lælius, Brutus, sive de Senectute, de Amicitia, de Claris Oratoribus Dialogi. With Notes by W. BROWNRIGG SMITH M.A., F.R.G.S. 2s.

16. **Livy:** History of Rome. Notes by H. YOUNG and W. B. SMITH, M.A. Part 1. Books i., ii., 1s. 6d.

16*. ——— Part 2. Books iii., iv., v., 1s. 6d.
17. ——— Part 3. Books xxi., xxii., 1s. 6d.

19. **Latin Verse Selections,** from Catullus, Tibullus, Propertius, and Ovid. Notes by W. B. DONNE, M.A., Trinity College, Cambridge. 2s.

20. **Latin Prose Selections,** from Varro, Columella, Vitruvius, Seneca, Quintilian, Florus, Velleius Paterculus, Valerius Maximus Suetonius, Apuleius, &c. Notes by W. B. DONNE, M.A. 2s.

21. **Juvenalis** Satiræ. With Prolegomena and Notes by T. H. S. ESCOTT, B.A., Lecturer on Logic at King's College, London. 2s.

GREEK.

14. Greek Grammar, in accordance with the Principles and **Philo**logical Researches of the most eminent Scholars of our own day. By **Hans Claude Hamilton.** 1s. 6d.

15,17. Greek Lexicon. Containing all the Words in General Use, with their Significations, Inflections, and Doubtful Quantities. By **Henry R. Hamilton.** Vol. 1. Greek -English, 2s. 6d.; Vol. 2. English-Greek, 2s. Or the Two Vols. in One, 4s. 6d.: cloth boards, 5s.

14,15,17. Greek Lexicon (as above). Complete, with the **Grammar**, in One Vol., cloth boards, 6s.

GREEK CLASSICS. With Explanatory Notes in English.

1. Greek Delectus. Containing Extracts from Classical Authors, with Genealogical Vocabularies and Explanatory Notes, by H. Young. New Edition, with an improved and enlarged Supplementary Vocabulary, by John Hutchison, M.A., of the High School, Glasgow. 1s. 6d.

2, 3. Xenophon's Anabasis; or, The Retreat of the Ten Thousand. Notes and a Geographical Register, by H. Young. Part 1. Books i. to iii., 1s. Part 2. Books iv. to vii., 1s.

4. Lucian's Select Dialogues. The Text carefully revised, with Grammatical and Explanatory Notes, by H. Young. 1s. 6d.

5-12. Homer, The Works of. According to the Text of **Baeumlein.** With Notes, Critical and Explanatory, drawn from the best and latest Authorities, with Preliminary Observations and Appendices, by T. H. L. Leary, M.A., D.C.L.

The Iliad: Part 1. Books i. to vi., **1s. 6d.** | Part 3. Books xiii. to xviii., 1s. 6d
                   Part 2. Books vii.to xii., **1s. 6d.** | Part 4. Books xix. to xxiv., 1s. 6d.
The Odyssey: Part 1. Books i. to vi., **1s. 6d.** | Part 3. Books xiii. to xviii., 1s. 6d.
                   Part 2. Books vii.to xii., **1s. 6d.** | Part 4. Books xix. to xxiv., and Hymns, 2s.

13. Plato's Dialogues: The Apology of Socrates, the Crito, and the Phædo. From the Text of C. F. Hermann. Edited with Notes, Critical and Explanatory, by the Rev. James Davies, M.A. 2s.

14-17. Herodotus, The History of, chiefly after the Text of **Gaisford.** With Preliminary Observations and Appendices, and Notes, Critical and Explanatory, by T. H. L. Leary, M.A., D.C.L.
Part 1. Books i., ii. (The Clio and Euterpe), 2s.
Part 2. Books iii., iv. (The Thalia and Melpomene), 2s.
Part 3. Books v.-vii. (The Terpsichore, Erato, and Polymnia), 2s.
Part 4. Books viii., ix. (The Urania and Calliope) and Index, 1s. 6d.

18. Sophocles: Œdipus Tyrannus. Notes by H. Young. 1s.

20. Sophocles: Antigone. From the Text of Dindorf. Notes, Critical and Explanatory, by the Rev. John Milner, B.A. 2s.

23. Euripides: Hecuba and Medea. Chiefly from the Text of Dindorf. With Notes, Critical and Explanatory, by W. Brownrigg Smith, M.A., F.R.G.S. 1s. 6d.

26. Euripides: Alcestis. Chiefly from the Text of Dindorf. With Notes, Critical and Explanatory, by John Milner, B.A. 1s. 6d.

30. Æschylus: Prometheus Vinctus: The Prometheus Bound. From the Text of Dindorf. Edited, with English Notes, Critical and Explanatory, by the Rev. James Davies, M.A. 1s.

32. Æschylus: Septem Contra Thebes: The Seven against Thebes. From the Text of Dindorf. Edited, with English Notes, Critical and Explanatory, by the Rev. James Davies, M.A. 1s.

40. Aristophanes: Acharnians. Chiefly from the Text of C. H Weise. With Notes, by C. S. T. Townshend, M.A. 1s. 6d.

41. Thucydides: History of the Peloponnesian War. Notes by H. Young. Book 1. 1s. 6d.

42. Xenophon's Panegyric on Agesilaus. Notes and Introduction by Ll. F. W. Jewitt. 1s. 6d.

43. Demosthenes. The Oration on the Crown and the Philippics. With English Notes. By Rev. T. H. L. Leary, D.C.L., formerly Scholar o Brasenose College, Oxford. 1s. 6d.

www.ingramcontent.com/pod-product-compliance
Lightning Source LLC
Chambersburg PA
CBHW030623030726
47497CB00006B/1609